Applications and Optimizations of Kalman Filter and Their Variants

Edited by Asadullah Khalid, Arif Sarwat and Hugo Riggs

Published in London, United Kingdom

Applications and Optimizations of Kalman Filter and Their Variants
http://dx.doi.org/10.5772/intechopen.1001639
Edited by Asadullah Khalid, Arif Sarwat and Hugo Riggs

Contributors
Abd El Mageed Hag Elamin Khalid, Akira Murakami, Alexander Chernodarov, Alwyn J. Hoffman, Arif Sarwat, Arya Abdolahi, Asadullah Khalid, Branko Kovačević, Eduardo A. Fernández, Hasan Iqbal, Hugo Riggs, Kazunori Fujisawa, Marco J. Suarez, Morteza Nazari-Heris, Navid Taghizadegan Kalantari, Oscar M. Sogamoso, Selma Cheshmeh Khavar, Takayuki Shuku, Tomislav Unkašević, Zoran Banjac

First published in London, United Kingdom, 2024 by IntechOpen
IntechOpen is the global imprint of INTECHOPEN LIMITED, registered in England and Wales, registration number: 11086078, 167-169 Great Portland Street, London, W1W 5PF, United Kingdom

British Library Cataloguing-in-Publication Data
A catalogue record for this book is available from the British Library

Additional hard and PDF copies can be obtained from orders@intechopen.com

Applications and Optimizations of Kalman Filter and Their Variants
Edited by Asadullah Khalid, Arif Sarwat and Hugo Riggs
p. cm.
Print ISBN 978-0-85466-566-2
Online ISBN 978-0-85466-565-5
eBook (PDF) ISBN 978-0-85466-567-9

For EU product safety concerns:
IN TECH d.o.o., Prolaz Marije Krucifikse Kozulić 3, 51000 Rijeka, Croatia,
info@intechopen.com or visit our website at intechopen.com.

Meet the editors

Dr. Asadullah Khalid received his bachelor's and M.S. degrees in electrical engineering from Aligarh Muslim University, India, and Illinois Institute of Technology, respectively, and his Ph.D. in electrical and computer engineering from Florida International University (FIU). He has been working in the consumer electronics and automotive industries for over nine years in different capacities. His research interests include battery management systems, microgrids, applied machine learning, and energy storage cybersecurity. He has published over 20 articles in various conferences and journals and has multiple registered U.S. patents.

Dr. Arif Sarwat has been in the industry (Siemens) and academia for more than 20 years. He is an eminent scholar chaired professor in the Department of Electrical and Computer Engineering and the director of the FPL-FIU Solar Research Facility at FIU (Florida International University). His research interests include smart grids, electric vehicles, high penetration renewable systems, storage and battery management systems, grid resiliency, large-scale data analysis, artificial intelligence, advanced metering infrastructure, smart city infrastructure, and cybersecurity. He has published more than 200 peer-reviewed articles and holds multiple patents. He currently has multiple projects funded by the National Science Foundation (NSF), industry, and the Department of Energy (DOE). This has received the NSF CAREER Award. He is the co-lead of the Masters in Energy & Cybersecurity education program. Previously, Dr. Sarwat worked at Siemens for more than nine years, winning three recognition awards. He is the co-author of a publication that won the Best Paper award at the Resilience Week in 2017 and a technical article that won both the Best Paper award in 2016 as well as the Most Cited Paper award in 2018 from Springer's Journal of Modern Power Systems and Clean Energy (MPCE). His team won the second-best paper award at the IEEE NAPS conference in the year 2020. Dr. Sarwat received the Faculty Award for Excellence in Research & Creative Activities in 2016, College of Engineering & Computing Worlds Ahead Performance in 2016, and FIU TOP Scholar Award in 2015 and 2019. He has been the chair of the IEEE Miami Section VT and Communication since 2012. He is an associate editor of the journal ACM Computing Surveys. For further information, please visit http://www.eps.fiu.edu.

Dr. Hugo Riggs has earned his Ph.D. in electrical and computer engineering from Florida International University, USA. He is an innovative researcher in the field of machine learning applications for photovoltaics and distribution system monitoring and has registered a patent on renewables forecasting technology.

Contents

Preface

In Applications and Optimizations of Kalman Filter and Their Variants, we meticulously examine the multifaceted applications and adaptations of Kalman filters across diverse domains, delving into their nuanced functionalities and optimizations.

The discussion begins with an analytical focus on the use of Kalman filters within microgrid management—a domain fraught with intricate challenges and presenting many technical opportunities for optimization. Within this context, the Kalman filter drives a grid-injected current for harmonic mitigation, offering adaptive and robust estimation techniques beneficial for effective microgrid control strategies.

The book continues with a study of distribution system state estimation in data collection, which is for system monitoring and control. To address the challenges posed by uncertain parameters such as the variable output of distributed generation, random meter errors, and inaccurate network parameters, an innovative two-stage stochastic programming model is developed using an extended Kalman filter to run state estimation with micro-phasor measurement unit data and attain a low error with minimal installation cost.

In later sections, the scope is expanded further to encompass pairs trading—a domain necessitating the fusion of financial modeling and Kalman filter-based estimation. From pairs trading to partial co-integration, each variation offers meticulous insights into the intricate interplay between price dynamics and fundamental value.

Kalman filter optimizations and variants are covered in a section that addresses the reduction of error in object tracking and thick-tailed Gaussian environments. It highlights the use of dynamic stochastic approximation algorithms, the minimum variance criterion, and the combination of M-robust estimates with minimum mean square error prediction. Additionally, it introduces the integration of a Huber moving window M-robust parameter estimator to handle unknown noise statistics, showcasing the effectiveness of the proposed approach in mitigating various types of outliers in a simulated maneuvering target tracking scenario.

Through numerous studies and rigorous analysis, we illuminate pathways toward enhanced efficiency and accuracy in Kalman filter applications, applying the powerful approach in a wide range of domains.

Dr. Asadullah Khalid and Dr. Hugo Riggs
Postdoctoral Associate,
Department of Electrical and Computer Engineering,
Florida International University,
Miami, Florida

Dr. Arif Sarwat
Eminent Scholar Chaired Professor,
Director of FPL-FIU Solar Facility and Energy, Power,
Sustainability and Intelligence (EPSi),
Department of Electrical and Computer Engineering,
Florida International University,
Miami, Florida

Section 1

Implementations for Microgrid, Geotechnical and Surveillance Applications

Chapter 1

Kalman Filter-Based Harmonic Distortion Mitigation Technique for Microgrid Applications

Hasan Iqbal, Asadullah Khalid, Hugo Riggs and Arif Sarwat

Abstract

This work presents a Kalman filter-based harmonic mitigation technique that can be used to reduce the total harmonic distortion (THD) of grid-injected current by reducing the harmonic distortion. Throughout the power supply system, power electronic converters have become increasingly used in recent years, which has resulted in more reactive power and harmonics. In a power system, harmonics add to noise and lower power quality. A study in MATLAB/Simulink 2022b for improving the THD of renewable energy injected grid currents using seven-level packed u-cells (PUC7) has been undertaken to improve the grid-injected current THD with Kalman filter-based harmonic mitigation technique. With the Kalman filter-based harmonic mitigation techniques, the THD for the grid-injected current is improved from 6.32 to 2.49% when compared to the grid-injected current without the Kalman filter-based technique.

Keywords: Kalman filter, harmonic mitigation, inverter, microgrid, THD

1. Introduction

A microgrid design for a sustainable energy ecosystem must include renewable energy sources. It is vital to utilize renewable energy sources in order to ensure a truly sustainable and eco-friendly future. Renewable energy sources such as solar and wind have become increasingly important with the demand for cleaner and more efficient electricity generation. This boom has also led to increasing importance of inverters as the key to converting fluctuating direct current (DC) from renewable sources into alternating current (AC) for grid integration. As renewable resources are converted into useful electrical power through inverters, this change is facilitated. When it comes to reducing our carbon footprint and switching to a greener energy source, understanding the role inverters play is crucial. Power from renewable sources can be variable and sometimes surprising, such as wind turbines and solar panels. Inverters serve as the main component of these systems, which convert direct current (DC) generated by renewable energy sources like solar panels and wind turbines into alternating current (AC), the type of electricity that people use in their homes and businesses [1, 2].

However, the switch to inverters has not been without its challenges. Inverters that incorporate renewable energy sources into the power system cause harmonic disturbances. It is mainly due to the switching activity that is inherent in power electronics that harmonic distortions appear as deviations from the ideal sinusoidal waveform of AC power. As a result of harmonic distortion, also known as total harmonic distortion (THD), connected loads are susceptible to being dependent and ineffective [3].

AC voltages and currents deviate from sinusoidal waveforms when they are subject to total harmonic distortion (THD). While converting DC power into AC power, power electronics inverters produce harmonics due to their non-linear nature. There are non-sinusoidal harmonics that can interfere with electrical equipment and impair sensitive loads. The reduction of harmonic distortion in microgrids has been accomplished using a wide range of techniques. One way to absorb harmonic currents is through passive harmonic filters. There may be some applications for which these filters are not appropriate, and they can be extremely expensive and large. A secondary strategy for canceling harmonics is active harmonic compensation, which involves injecting extra currents into the power supply proactively. Often, these strategies require extra hardware and intricate control algorithms, which can be implemented with digital signal processors or harmonic controllers [4, 5].

As harmonic distortions become more prevalent, power electronics researchers are exploring several mitigation strategies. THD and power quality have been improved by using active and passive filters, as well as sophisticated control algorithms. The continuous search for the best solution involves weighing the advantages and disadvantages of all approaches [6].

Many harmonic reduction techniques are used in the industry, including phase multiplication, active filtering, hybrid filters, passive filters, and so on. In the past, passive power filters were widely used to reduce harmonic distortion. They did this by absorbing harmonic currents and simplifying and streamlining filter construction. An impedance level is specified to remove a particular harmonic. However, the passive filter has several drawbacks, including a high configuration size, fixed harmonic correction, and the creation of resonance with the power system network. Active power filters (APFs), the most popular and well-respected harmonic mitigation filters in the market, can be utilized to mitigate the shortcomings of passive power filters [7, 8].

As opposed to passive power filters, active power filters use active components such as resistors, capacitors, and inductors, while passive power filters use passive components such as resistors, capacitors, and inductors. A power filter can be categorized into two categories, namely shunt or series, based on the type of active component. There is a difference between a shunt filter and a series APF in the sense that the load and filter are connected in parallel, whereas the load and filter in a shunt filter are connected in series [9]. Shunt active filters produce sinusoidal current waveforms by applying harmonic compensating current that is equal in magnitude and opposite in phase to the harmonics in the current. As opposed to the shunt active filter, which reduces distortions associated with currents, the series active filter reduces distortions associated with voltages [7, 8].

This study presents the Kalman filter-based strategy, a powerful harmonic mitigation method for microgrid applications. With the use of the Kalman filter, a popular recursive technique for estimation and control, this methodology seeks to improve power quality in microgrid contexts by dynamically mitigating harmonic distortions in real time. Through an examination of the fundamentals of Kalman filters and their use in harmonic mitigation, this research aims to further the field of cutting-edge technologies for more dependable and cleaner microgrid operations.

The remaining chapters are arranged as follows: Sections 2 and 3 provide the literature reviews for the various inverters and Kalman filters, respectively. In Part 4, the design and modeling of the power electronics converter for renewable energy using Kalman filters-based harmonic mitigation is completed. In Part 5 and Section 6, the simulation results are finally examined and concluded.

2. Inverters and harmonic distortions

Multi-level inverters (MLIs) were developed in the 1970s to address the shortcomings of two-level inverters, such as excessive harmonics. But this required the system to have additional power semiconductors put to it. When the price of semiconductors dropped, this was a fair trade-off. Numerous undesirable characteristics are associated with typical two-level converters, including bulky filters, high switch stress, significant distortion caused by harmonics in the voltage waveforms, and restricted power handling capacity. It is not recommended to utilize two-level converters in grid-connected and particularly high-powered applications, since they are inappropriate for use in a range of usage, including all of the ones listed above [10, 11].

The use of multi-level inverters (MLI) has enhanced reliability, allowed power switches to be rated above their capabilities, and obtained high-quality waveforms. They may now be utilized in high-power applications because of this. In the past several decades, a number of publications pertaining to control and modulation approaches have been published. In practice, three types of MLI have been studied most often: cascaded H-bridge converters (CHBs), floating capacitor converters (FCs), and neutral point clamped converters (NPCs). For high-power applications, the scalable and modular CHB architecture is preferred. The drawback of CHB is that many independent voltage sources are needed. When there are additional levels, the FC and NPC structures need more capacitors, and the CHB structure needs more DC sources because it has H-bridges [12–14].

The most common topology is the CHB topology due to its modular design, ease of usage, and need for a lesser count of components. The AC output voltage of the CHB is staircase sinusoidal as a result of the utilization of several DC sources. In this work, the topology utilized is the seven-level packed U-cell (PUC7), which is shown in **Figure 1** [15, 16]. The PUC7 utilizes six switches, one source, and one capacitor balanced at one-third of the DC source to obtain a seven level of inverter output

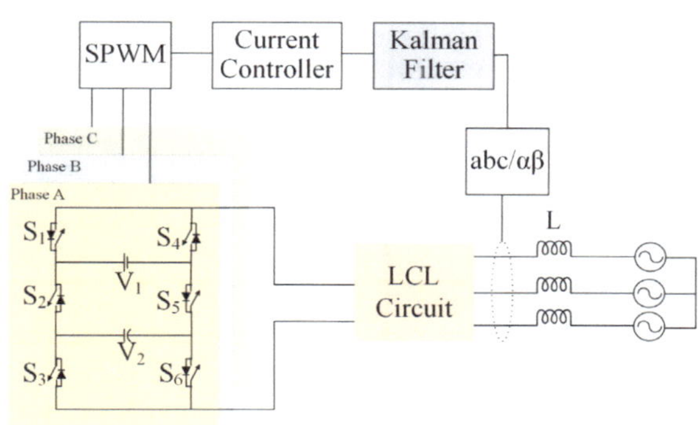

Figure 1.
Packed U-cell topology with Kalman filter.

Switching states	S_1	S_2	S_3	V_{ab}	Capacitor voltage
1	1	0	0	V_1	—
2	1	0	1	$V_1 - V_2$	↑
3	1	1	0	V_2	↓
4	1	1	1	0	—
5	0	0	0	0	—
6	0	0	1	$-V_2$	↓
7	0	1	0	$V_2 - V_1$	↑
8	0	1	1	$-V_1$	—

"↑"—charging, "—"—no effect, "↓"—discharging.

Table 1.
Switching table for PUC7.

voltage. The switching schemes for this topology are presented in **Table 1**. In this table, capacitor voltage across each state is also tabulated.

Certain controlling techniques are employed in order to improve the output power and functioning of modulating index (MI). Based on regulating and modulating approaches, MI operates as either high or fundamental switching frequency modulation (FSFM). Switch power loss is substantial when the MI is operated at a high switching frequency. A FSFM approach minimizes the loss of power due to switching in high-power networks.

There are three types of FSFM: Selective harmonic elimination (SHE), closest level control (NLC), and nearest space vector control (NSVC). Using the NSVC to find the space vector is challenging and time-consuming. Despite its low modulation index and low output level, the NLC scheme produces a great deal of harmonic distortion. Moreover, these regulating techniques cannot eliminate the particular harmonics. Up to 1 kHz, low switching frequency SHE is suitable. SHE techniques are used with optimal switching angles selected to eliminate the distinct lower-order harmonics. A selective harmonic elimination modulation based on the crystal structure algorithm (CryStAl) is employed in cascaded H-bridge MLIs. Research has shown that CHB architecture is the most widely used architecture, as it eliminates the need to balance floating capacitor voltage and account for the DC side's midpoint voltage offset.

The use of MLIs can benefit all kinds of power applications, including drives, storage, and FACTS tools. There are several advantages to using MLIs over conventional inverters, including higher output efficiency, improved electromagnetic compatibility, a lower rating, a sine wave output, a switch that is near to the ground, and fewer tools needed to generate higher voltages with better power quality when compared to conventional inverters. Currently, inverters are being designed with fewer components, such as DC sources and switches, to minimize their size. A total harmonic distortion (THD) standard specified by IEEE Standard 519-2022 must be adhered to when calculating converter output voltage. With more stages/levels, the THD diminishes, improving the power quality.

Harmonic distortion (THD) is a key component of an efficient inverter. According to IEEE standards, inverters can reduce total harmonic distortion by employing selective harmonic mitigation (SHM) because the overall harmonic distortion is reduced.
The output waveform of the inverter in **Figure 2** is analyzed using Fourier series

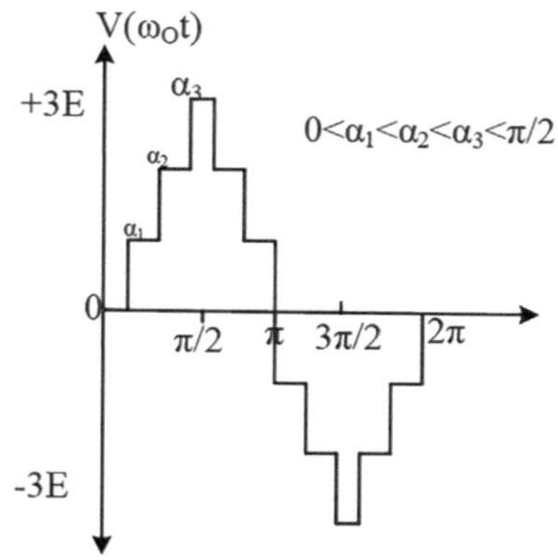

Figure 2.
Seven-level output voltage with switching angles.

analysis to obtain SHM Equations. An output voltage with seven levels can be written as follows:

$$\text{V(t)} = \text{E}\left[\sum_{i=1}^{3} u(t-\alpha_i) - \sum_{i=1}^{3} u(t-(\pi-\alpha_i)) - \sum_{i=1}^{3} u(t-(\pi+\alpha_i)) + \sum_{i=1}^{3} u(t-(2\pi+\alpha_i))\right] \tag{1}$$

Because of its symmetry on the x-axis and the y-axis, the waveform's Fourier series contains odd harmonics whose amplitudes are as follows:

$$\text{V}_n = \frac{2}{T_o}\int_0^{t_0} V(t)\sin(n\omega_o t)dt \forall n = 1, 3, 5, \ldots \ldots \tag{2}$$

Replacing V(t) from Eqs. (1) into (2), gives to

$$\text{V}_n = \frac{4}{n\pi}\left[\sum_{i=1}^{3} E\cos(n\alpha_i)\right] \forall n = 1, 3, 5, \ldots \ldots \tag{3}$$

where $0 < \alpha_1 < \alpha_2 < \alpha_3 < \pi/2$

$$\text{THD} = \frac{\sqrt{\sum_{n=3,5,\ldots}^{\infty}\left[\frac{4E}{n\pi}\sum_{i=1}^{3}\cos(n\alpha_i)\right]}}{\frac{4E}{\pi}\sum_{i=1}^{3}\cos(n\alpha_i)} \tag{4}$$

Optimally, using a fundamental voltage V_1, one may compute the switching angles $\alpha_1, \alpha_2, \alpha_3$ to obtain $V_0(\omega t) = V_1 sin\,(\omega t)$, which will result in a null high-order harmonic content. In order to minimize n harmonics and manage the fundamental output voltage, $n+1$ equations are required. The SHE switching angles can be obtained by simultaneously solving the following equations:

$$\left.\begin{array}{l}\cos(\alpha_1) + \cos(\alpha_2) + \cos(\alpha_3) = \dfrac{4mE}{\pi} \\ \cos(3\alpha_1) + \cos(3\alpha_2) + \cos(3\alpha_3) = 0 \\ \cos(5\alpha_1) + \cos(5\theta_2) + \cos(5\alpha_3) = 0 \\ . \\ . \\ . \\ \cos(n\alpha_1) + \cos(n\alpha_2) + \cos(n\alpha_3) = 0\end{array}\right\} \tag{5}$$

3. Kalman filters: theory and applications

Kalman filters are used in state space format to identify states based on linear dynamical systems. The state's evolution from instant $K-1$ to instant k is defined as follows in accordance with the process model:

$$x_k = Fx_{k-1} + Bu_{k-1} + w_{k-1} \tag{6}$$

where w_{k-1} is the process noise vector, B is the control-input matrix applied to the control vector u_{k-1}, and F is the state transition matrix applied to the prior state vector x_{k-1}.

The measurement model, paired with the process model, explains the correlation between state and value at present time step k as follows:

$$z_k = Hx_k + v_k \tag{7}$$

Here is the equation: z_k = measurement vector, H = measurement matrix, v_k = Noise vector.

The two phases of a Kalman filter are prediction and update. Note that the phrases "prediction" and "update" are occasionally described to as "propagation" and "correction" in numerous literary works, respectively. An overview of the Kalman filter algorithm is provided below:

Prediction stage:

Predicted state estimation

$$\hat{x}_k^- = F\hat{x}_{k-1}^+ + Bu_{k-1} \tag{8}$$

Predicted error covariance

$$P_k^- = FP_{k-1}^+F^T + Q \tag{9}$$

Updating stage:

Measurement residual

$$\tilde{y}_k = z_k - H\hat{x}_k^- \tag{10}$$

Kalman gain.

$$K_k = P_k^- H^T\left(R + HP_k^-\right) \tag{11}$$

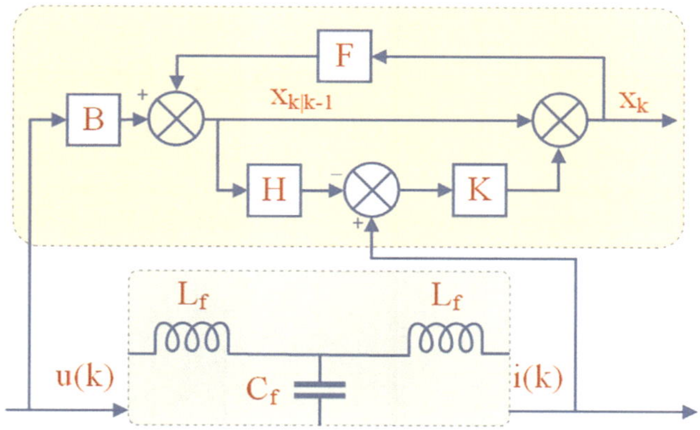

Figure 3.
Structure of the Kalman filter.

Updated state estimate

$$\hat{x}_k^+ = \hat{x}_k^- + K_k \tilde{y} \tag{12}$$

Updated error estimate

$$P_k^+ = (I - K_k H) P_k^- \tag{13}$$

In the above equations, the estimate of a variable is represented by ^ operator and the priori and posteriori are represented by + and – superscripts.

4. Designing a Kalman filter for harmonic mitigation

The optimal estimation recursive data processing algorithm known as the Kalman filter is used to infer a system's state from erratic observations. It is used in domains such as computer vision, econometrics, guidance, navigation, and signal processing. In order to minimize harmonics, this work calculates voltage and current magnitudes as shown in **Figure 3** using the Kalman filter [9]. It operates using a prediction-updating methodology. The Kalman filter algorithm's dynamical and measurement equations are provided by

$$U(k+1) = FU(k) + Bh(k) \tag{14}$$

$$W(k) = HU(k) \tag{15}$$

The state vector can be described as:

$$U(k) = [i_{i1}(k) \;\; i_{o2}(k) \;\; i_{o1}(k)]^T \tag{16}$$

where, in turn, i_{i1}, i_{o2}, and i_{o1} stand for input phase current (phase −1), output phase voltage (phase −2), and output phase current (phase −1) which can be initiated as $[0\ 0\ 0]^T$. The input-containing vector is shown as

$$h(k) = [v_l(k) \quad v_{dc}(k) \quad v_{o1}(k)]^T \tag{17}$$

where the output line voltage, inverter output voltage, and output phase-1 voltage are represented by the variables v_l, v_{dc}, and v_{o1}, respectively. Process noise vector $s(k)$ and measurement noise vector $r(k)$ are represented by these vectors. The observation matrix, H chooses the measured variables of the states vector, and $W(k)$ is the measurement vector. Every parameter in $U(k)$ has an impact on the system's state, which is applied by the state transition matrix F. The matrix F is provided by

$$F = \begin{bmatrix} 1 & \frac{-T}{L_f} & 0 \\ \frac{T}{C_f} & 1 & \frac{-\mathrm{T}}{C_f} \\ 0 & \frac{T}{L_f} & 1 \end{bmatrix} \tag{18}$$

where T represents the sampling duration and L_f and C_f indicate the filters' capacitance and inductance, respectively. The input control matrix, denoted by B, employs the effects of every constraint in $h(k)$ to the system's state. The matrix B is provided by

$$B = \begin{bmatrix} \frac{T}{L_f} & \frac{-T}{L_f} & 0 \\ 0 & 0 & 0 \\ 0 & 0 & \frac{-T}{L_f} \end{bmatrix} \tag{19}$$

There are two parts to the Kalman filter algorithm, namely the prediction part and the updating part.

Part 1: The prediction part of this model is responsible for predicting the a priori estimate $\hat{U}^-(k)$ and error covariance for the a priori estimate $Z^-(k)$.

$$\hat{U}^-(k) = F\hat{U}(k-1) + Bh(k) \tag{20}$$

$$Z^-(k) = FZ(k-1)F^T \tag{21}$$

Part 2: The Kalman gain, $K(v)$, is used to determine the posteriori estimate, $\hat{U}(k)$. To reduce the a posteriori estimate error covariance, $Z(k)$, the Kalman gain is calculated.

$$K(k) = Z^-(k)H^T\left(HZ^-(k)H^T\right)^{-1} \tag{22}$$

$$\hat{U}(k) = \hat{U}^-(k) + K(k)\left(W(k) - H\hat{U}^-(k)\right) \tag{23}$$

$$Z(k) = (I - K(k)H)Z^-(k) \tag{24}$$

The identity matrix is represented by I. Using Eqs. (20) and (23) yields the recursive posteriori estimate equation, which is represented by

$$\hat{U}(k) = \left(F\hat{U}(k-1) + Bh(k)\right)(I - K(k)H + K(k)W(k)) \tag{25}$$

A priori error covariance can be estimated using Eq. (24) in Eq. (21) as follows:

$$Z^{-}(k) = F(I - K(k-1)H)Z^{-}(k-1)H^{T} \tag{26}$$

The switching action of the inverter power transistor in the active filter circuit can be controlled by selective harmonic mitigation by the output of the Kalman filter block. The state vector serves as the input for the Kalman filter block in the active filter circuit to control the switching action of the Kalman filter block.

5. Results

The simulation results are discussed in this section in order to give you a better understanding of them. Kalman filter-based harmonic mitigation for grid-tied PUC7 inverters is implemented using MATLAB/Simulink® 2022b in a MATLAB/Simulink® environment with parameters shown in **Table 2**.

Particular	Values
DC source (V_1)	315 V
Capacitor (C_2)	8000 μF
Line impedance	0.2 Ω and 2.5 mH
AC grid voltage (peak)	220 V
Line frequency	60 Hz
Sampling time	10 μs
Initial covariance (P)	Identity matrix (3 × 3)
Process noise covariance (Q)	0.05
Measurement noise covariance (R)	0.01

Table 2.
Simulation parameters.

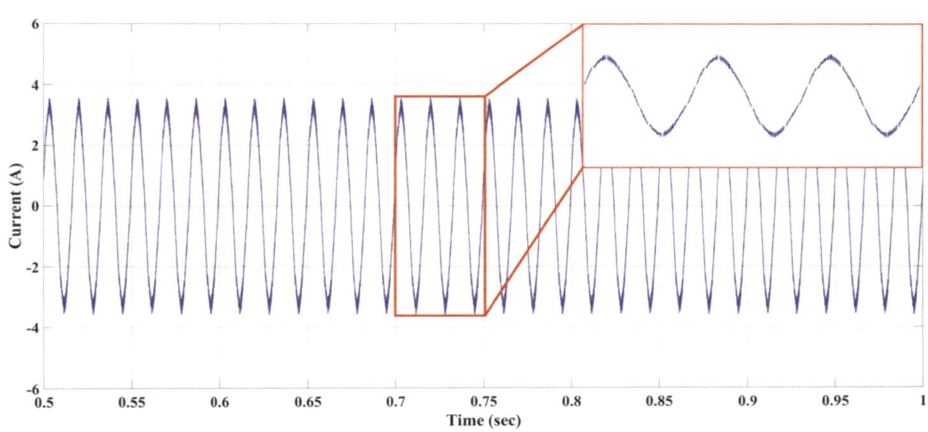

Figure 4.
Grid-injected current without Kalman filter.

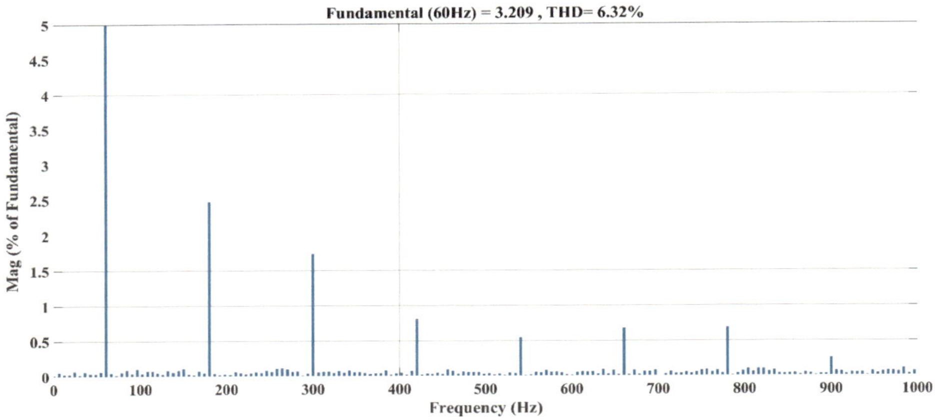

Figure 5.
FFT analysis of the grid-injected current without Kalman filter.

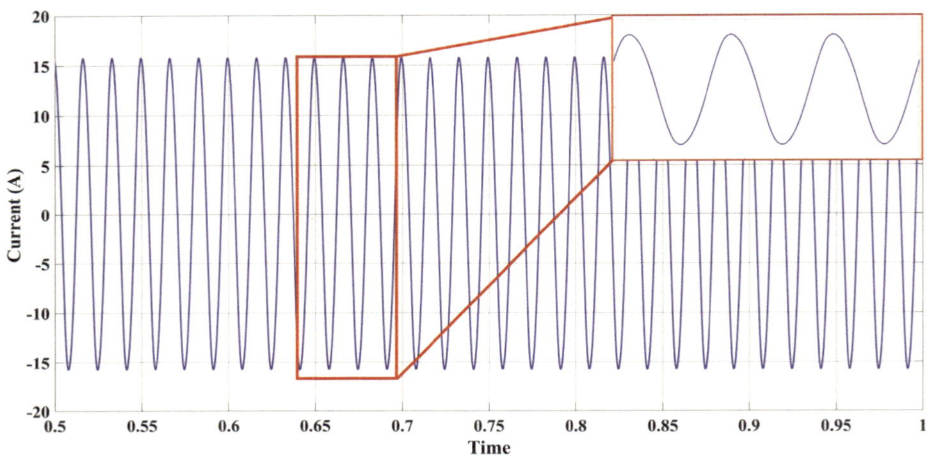

Figure 6.
Grid-injected current with Kalman filter.

In **Figure 4**, the grid-injected current for the inverter output without Kalman filter is plotted and **Figure 5** shows the FFT analysis for the same. The Kalman filter-based mitigation is applied and the grid current is obtained in **Figures 6** and 7 shows the FFT analysis for the grid-injected current for the inverter. The THD without Kalman filter-based technique was 6.42% which is improved to 2.49% with the proposed Kalman-based technique.

The comparison of the proposed Kalman filter-based techniques with other filters is tabulated in **Table 3**.

6. Conclusions

To reduce the harmonic distortion of the inverter, a Kalman filter-based harmonic mitigation strategy is used. The grid is not healthy because harmonic distortion is

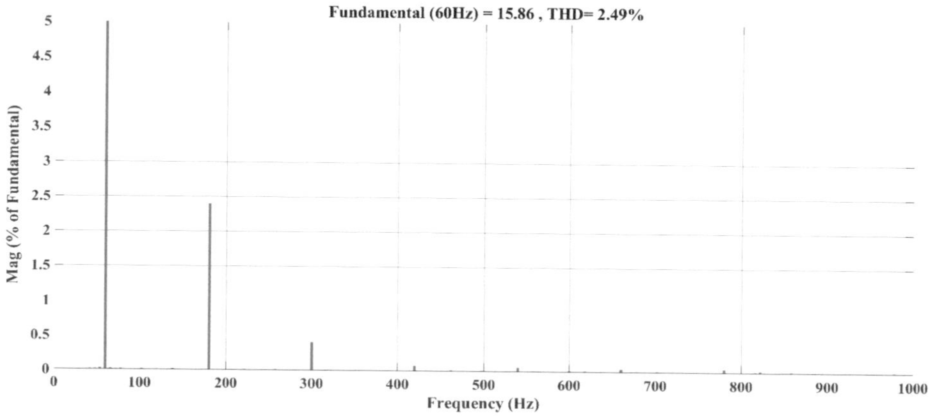

Figure 7.
FFT analysis of the grid-injected current with Kalman filter.

Particular	Current harmonic content (%)
Without filter	34
Passive filter	20
Proposed Kalman Filter	2.49

Table 3.
Comparison with other existing filters.

created by the presence of power converter components. Due to the reduction of harmonics, the output current power quality has been significantly improved. By generating a harmonic voltage with a similar amplitude but opposite phase, the filter compensates for voltage harmonics, resulting in sinusoidal voltage and current waveforms. In compliance with IEEE 519-2022 standards, the THD was 6.32% prior to the suggested Kalman filter and decreased to 2.49% following the Kalman filter-based mitigation measures. Increasing the number of renewable energy sources connected to the grid will result in power converters being used more frequently to convert DC power to AC power, resulting in harmonic injected current. Since these harmonics are detrimental to the system's functionality, suitable power converters and control schemes must be developed for renewable energy that is connected to the grid.

Conflict of interest

The authors declare no conflict of interest.

Author details

Hasan Iqbal, Asadullah Khalid, Hugo Riggs and Arif Sarwat*
Electrical and Computer Engineering, Florida International University, Miami, USA

*Address all correspondence to: asarwat@fiu.edu

References

[1] Rodriguez J, Lai J-S, Peng FZ. Multilevel inverters: A survey of topologies, controls, and applications. IEEE Transactions on Industrial Electronics. 2002;**49**(4):724-738. DOI: 10.1109/TIE.2002.801052

[2] Bughneda A, Salem M, Richelli A, Ishak D, Alatai S. Review of multilevel inverters for PV energy system applications. Energies. 2021;**14**(6):1585. DOI: 10.3390/EN14061585

[3] Lei Y et al. A 2-kW single-phase seven-level flying capacitor multilevel inverter with an active energy buffer. IEEE Transactions on Power Electronics. 2017;**32**(11):8570-8581. DOI: 10.1109/TPEL.2017.2650140

[4] Nirmal Mukundan CM et al. A new multilevel inverter based grid connected reliable solar power transfer unit with power quality enhancement. IEEE Transactions on Industry Applications. 2023;**59**(2):1887-1900. DOI: 10.1109/TIA.2022.3218523

[5] Pourfarrokh S, Adabi J, Zare F. A novel multilevel inverter with self-balancing capability of capacitors voltage; structure, modulation and operation. IEEE Journal of Emerging and Selected Topics in Power Electronics. 2023;**11**(2):1854-1864. DOI: 10.1109/JESTPE.2022.3222344

[6] Roy S, Debnath A, Tariq M, Behnamfar M, Sarwat A. Characterizing current THD's dependency on solar irradiance and supraharmonics profiling for a grid-tied photovoltaic power plant. Sustainability. 2023;**15**(2):1214. DOI: 10.3390/SU15021214

[7] Luo S, Wu W, Koutroulis E, Chung HSH, Blaabjerg F. A new Kalman-filter-based harmonic current suppression method for the virtual oscillator controlled grid-tied inverter. IEEE Journal on Emerging and Selected Topics in Circuits and Systems. 2022;**12**(1):251-259. DOI: 10.1109/JETCAS.2022.3141106

[8] Chen B, Wu W, Gao N, Yao Z, Chung H, Blaabjerg F. Kalman-filter-estimation based sliding mode control of three-phase LCL-filtered grid-tied inverter using only grid-injected current sensors. In: IEEE 9th International Power Electronics and Motion Control Conference (IPEMC2020- ECCE Asia). 2020. pp. 2368-2373. DOI: 10.1109/IPEMC-ECCEAsia48364.2020.9367829

[9] Ravikumar A, Mohan N, Soman KP. Performance enhancement of a series active power filter using Kalman filter based neural network control strategy. In: 2018 International Conference on Advances in Computing, Communications and Informatics (ICACCI). 2018. pp. 1702-1706. DOI: 10.1109/ ICACCI.2018.8554889

[10] Dewangan P, Dewangan PK, Nagdeve UT. A review of single-phase grid-connected inverters for photovoltaic generation system. International Journal of Research. 2014; **1**(9):521-525. DOI: 10.1109/TIA.2005.853371

[11] Akbari A, Poloei F, Bakhshai A. A brief review on state-of-the-art grid-connected inverters for photovoltaic applications. In: IEEE International Symposium on Industrial Electronics. 2019. pp. 1023-1028. DOI: 10.1109/ISIE.2019.8781166

[12] Nabae A, Takahashi I, Akagi H. A new neutral-point-clamped PWM inverter. IEEE Transactions on Industry

Applications. 1981;**IA-17**(5):518-523. DOI: 10.1109/TIA.1981.4503992

[13] Peng FZ. Z-source inverter. In: IEEE Trans- actions on Industry Applications. Vol. 39, no. 2. 2003. pp. 504-510. DOI: 10.1109/TIA.2003.808920

[14] Gupta KK, Ranjan A, Bhatnagar P, Sahu LK, Jain S. Multilevel inverter topologies with reduced device count: A review. IEEE Transactions on Power Electronics. 2016;**31**(1):135-151. DOI: 10.1109/TPEL.2015.2405012

[15] Iqbal H, Tariq M, Sarfraz M, Anees MA, Alhosaini W, Sarwar A. Model predictive control of packed U-cell inverter for microgrid applications. Energy Reports. 2022;**8**: 813-830. DOI: 10.1016/j.egyr.2022. 05.188

[16] Pakdel M, Jalilzadeh S. A new family of multilevel grid connected inverters based on packed U cell topology. Scientific Reports. 2017;7(1):12396. DOI: 10.1038/s41598-017-12806-5

Chapter 2

An Application of Particle Filter for Parameter Estimation and Prediction in Geotechnical Engineering

Akira Murakami, Takayuki Shuku and Kazunori Fujisawa

Abstract

This chapter presents an application example of a nonlinear Kalman Filters (KFs), i.e., Particle Filter (PF), for state (or parameters) estimation and prediction of a dynamical system in geotechnical engineering. First key characteristics of dynamical systems in geotechnics, which need to be considered in filtering, are described by showing some figures, and why PF is necessary for geotechnical applications is explained. Then, a detailed algorithm and implementation of PF for geotechnical problems are presented with key equations. The PF is demonstrated through a case history focusing on deformation behavior of a ground due to embankment construction. The PF is applied to estimation of geotechnical parameters and predictions of future settlement behavior of the ground to discuss the effectiveness of the PF in geotechnical engineering. The results of the case history have shown that PF has presented great promise as an accurate parameter identification for a nonlinear dynamic model. The simulation with the identified parameters predicts the actual measurement data with high accuracy even though a limited amount of measurement data was used in identification stage. The PF provides more information on estimates than optimization methods because the estimates are obtained in the form of probability density functions (PDFs). This characteristic can contribute to risk analysis and reliability-based decision-making in geotechnical practice.

Keywords: nonlinear Kalman filter, particle filter, geotechnical engineering, probability density function, finite element method

1. Introduction

Filtering refers to the process of estimating state or parameters and prediction of dynamic systems based on noisy monitoring data, and it has significant applications in many different fields. There is no exception in geotechnical engineering: there are many applications of filtering for geotechnical problems, particularly for enhancing the accuracy and reliability of numerical simulations. Filtering helps in fine-tuning numerical simulation models by continuously updating them as new data becomes available.

This iterative process enhances the accuracy and reliability of the models, leading to better decision-making in engineering projects. One of the key benefits of filtering is its ability to quantify uncertainties in model predictions. Geotechnical engineering involves dealing with uncertain soil properties, geological variations, and environmental conditions [1]. Filtering provides a probabilistic framework to express these uncertainties, which is crucial for risk assessment and management in engineering projects.

One of the most well-known methods or algorithms to achieve filtering is KF. In 1960, R.E. Kalman published a famous paper describing a recursive solution to the discrete-data linear filtering problem [2]. This algorithm has been applied to the *Apollo* program. Since that time, due in large part to advances in digital computing, the KF has been the subject of extensive research and application, particularly in the area of time-series analysis. The KF is a set of mathematical equations that provides an efficient computational recursive solution of the least-squares method (LSM). The filter is very powerful in several aspects: it supports estimations of past, present, and even future states, and it can do so even when the precise nature of the modeled system is unknown.

KF also received significant attention in geotechnical engineering around 1980s. Murakami and Hasegawa [3] applied KF to a geotechnical one-dimensional time-series problem (consolidation settlement problem) for updating geotechnical parameters and predicting consolidation settlement of soft clay ground. Murakami and Hasegawa [4, 5] extended their method to two-dimensional settings: they integrated KF and Finite Element Method (FEM), which is known as "Kalman Filter FEM", for solving broad practical/realistic time-series estimation problems in geotechnical engineering. In addition, they utilized it successfully to solve the optimal location of observation points [6, 7]. After these pioneering works, several researches combining KF with numerical simulation models were reported in geotechnical engineering [8–11].

Although the KF is a powerful tool for state (or parameter) estimations and prediction, it has some limitations which are as follows:

1. Linearity: Standard KF assumes that the system dynamics and measurement models are linear. This can be a significant limitation for systems that are inherently nonlinear.

2. Gaussianity: KF assumes that the process and measurement noise are both Gaussian. If the noise characteristics deviate significantly from a Gaussian distribution, the filter's performance may degrade.

3. Robustness: KF may not be robust to outliers or sudden changes in the system dynamics, which can lead to divergence or inaccurate estimates.

Unfortunately, the characteristics of geotechnical problems can make the application of the KF challenging. First, many geotechnical systems exhibit nonlinear behavior, such as soil-structure interaction and material nonlinearity (e.g., plasticity, strain-hardening). Second, geotechnical systems often encounter non-Gaussian noise due to complex subsurface conditions, measurement errors, and model uncertainties. KF assumes Gaussian noise, which may not be suitable for accurately capturing the uncertainties in geotechnical problems.

Despite these challenges, KF and its variants are still valuable tools in geotechnical engineering for applications such as deformation prediction, slope stability analysis, and foundation performance assessment. The variants include Extended Kalman Filter (EKF) [12] for nonlinear systems, the Unscented Kalman Filter (UKF) [13], Ensemble Kalman Filter (EnKF) [14] and Particle Filter (PF) [15–17] for better handling of nonlinearities, and adaptive versions for systems with changing dynamics. Due to the increased need for nonlinear filters in geotechnical engineering, several attempts have been made to apply these nonlinear filters for enhancing the accuracy and reliability of numerical simulations, managing uncertainties, adapting to changing conditions, and supporting cost-effective and safe engineering solutions. Its integration into geotechnical practices represents a significant advancement in the field, enabling more informed and confident decision-making in complex engineering projects [18–23].

This chapter describes some key characteristics of geotechnical problems that need to be considered in filtering and presents an application example of a variant of nonlinear KF for parameter estimation and predictions in geotechnical engineering. To the best of the authors' knowledge, PF is the most suitable method for geotechnical applications because of its versatility for nonlinear and non-Gaussian problems. We believe that this chapter is beneficial not only for geotechnical engineers and researchers but also for those who learn standard Kalman filtering and other nonlinear Kalman filtering. This chapter is organized as follows: Section 2 describes the importance and the roles of state estimations and predictions based on monitoring data in geotechnical engineering considering characteristics of geotechnical problems. Section 4 shows an application example of PF in a numerical simulation of a typical geotechnical problem: predicting deformation behavior of a ground due to embankment construction. Section 5 briefly summarizes this chapter.

2. Characteristics of geotechnical problems

Understanding heterogeneity of grounds and variability of soil parameters is crucial in geotechnical engineering. This section outlines the importance of state estimation and prediction based on monitoring data by considering heterogeneity of grounds and uncertainties in geotechnical parameters.

The concept of ground heterogeneity is a fundamental aspect of geotechnical engineering. It refers to the variation in the composition, structure, and properties of the ground, including soil and rock layers. This heterogeneity can occur on various scales and has significant implications for construction, environmental projects, and geological studies. At least two types of heterogeneity in the context of geotechnical engineering, which are a) soil composition: variation in the types of soil (clay, sand, silt, and gravel) and their mixtures, and b) rock structure: differences in rock types, stratification, faulting, folding, and weathering. **Figure 1** shows an example of heterogeneity of the ground created by a coupled Markov chain (CMC) method [24]. The heterogeneity can be due to natural processes like sedimentation, erosion, and weathering, or human activities such as excavation or filling. Therefore, the heterogeneity of the ground can be observed not only in natural grounds, but also artificial grounds such as earthen embankments for river dikes.

Variability (or uncertainty) of soil parameters is another fundamental aspect in geotechnical engineering that influences the design and analysis of geotechnical structures. These uncertainties arise due to the complex and variable nature of soil as a material, and they have significant implications for the safety, reliability, and cost-effectiveness of geotechnical projects. There are many potential sources of natural

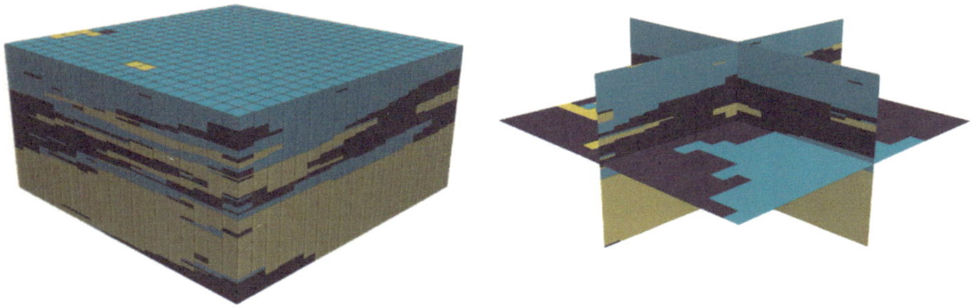

Figure 1.
An example of heterogeneity of a ground [24].

variability of soils, which are a) natural variability: soils are formed through long geological processes, leading to inherent variability in properties like grain size, mineral composition, density, strength, and permeability, b) sampling and testing limitations: soil properties are typically determined through field sampling and laboratory testing, but samples may not represent the entire site, and tests can have inherent errors or limitations, C) spatial variability: soil properties can vary significantly over short distances, both horizontally and vertically, making it challenging to obtain a comprehensive understanding of the subsurface conditions, and d) temporal changes: soil properties can change over time due to factors like weathering, organic matter decomposition, moisture content changes, and human activities. **Figure 2** shows examples of variability of rock parameters. We collected compressive strength and elastic modulus of granite from several technical reports published in Japan. It is clear that the parameters show large variations despite all the data being from Japanese granite.

The uncertainties dealt with in geotechnical engineering are classified as being either *aleatory* – the inherent variation given sufficient samples of the stochastic process, which can be characterized via a probability distribution, or *epistemic* – where

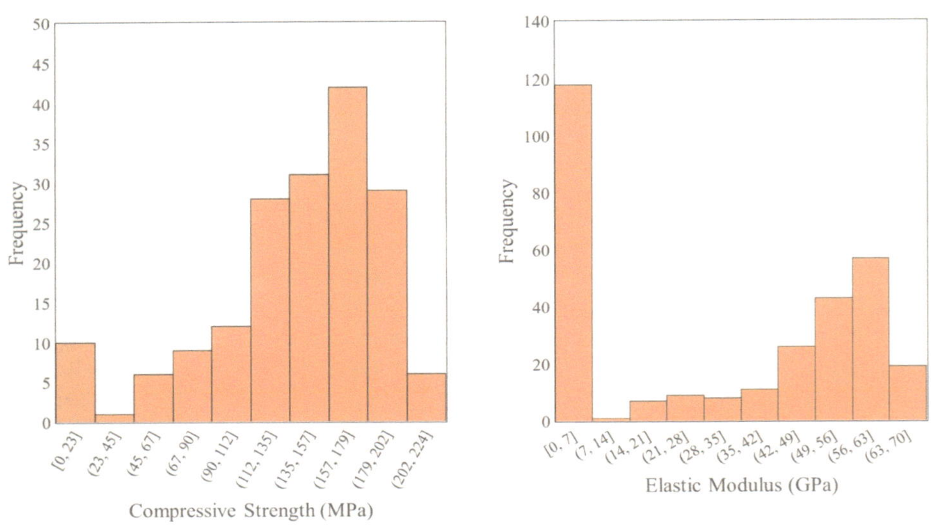

Figure 2.
Variability of rock properties.

there is insufficient information concerning the parameters of interest. The aleatory uncertainty is also called stochastic uncertainty or randomness. On the other hand, the epistemic uncertainty is referred to as knowledge uncertainty or reducible uncertainty. The categories of uncertainty in soil properties are shown in **Figure 3**. These uncertainties need to be considered in geotechnical engineering.

The heterogeneity and the parameter variability make numerical simulations in geotechnical engineering challenging. As shown in **Figure 1**, natural (and even artificial) grounds are essentially heterogeneous, and it is challenging to make numerical models considering the actual heterogeneity in simulations. Most of practical simulations assume that the ground is homogeneous for simplicity. In addition, geotechnical researchers and engineers usually assume that the input parameters are "deterministic" despite actual parameters being quite uncertain (**Figure 2**). Geotechnical engineers select one parameter from the dataset of the parameters and tend to choose "conservative" parameters to avoid the worst-case scenario.

Rather than using single, deterministic values for soil parameters and ground conditions, probabilistic approaches are necessary to deal with the uncertainties. In general, uncertainty quantification is necessary, as current real-world data is insufficient and incomplete. A renowned statistician C. R. Rao stated that the logical equation is important in decision-making under uncertainties (**Figure 4**). If we have to make decisions under uncertainty, we cannot avoid mistakes. If mistakes cannot be avoided, we had better know how often we make mistakes [26]. Therefore, it is natural to consider the probabilistic methods as suitable for geotechnical applications.

As mentioned in the previous section, one of the key benefits of Bayesian updating is its ability to quantify uncertainties. Uncertainties in geotechnical engineering can be quantitatively modeled with PDFs or statistical models, and the initial probability or

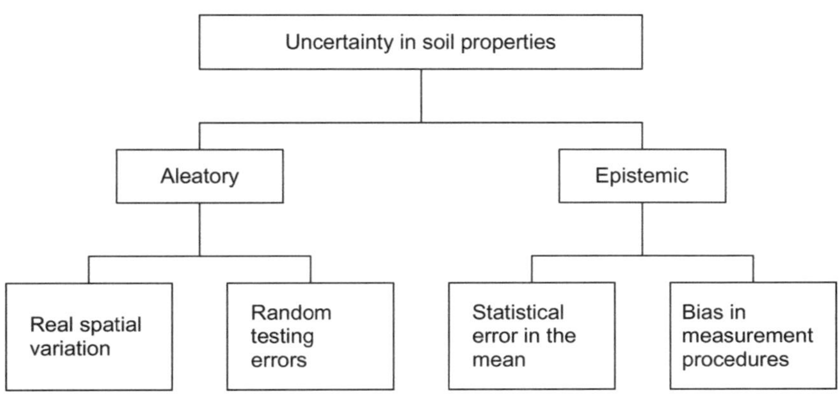

Figure 3.
Categories of uncertainty in soil properties [25].

Figure 4.
A logical equation stated by C.R. Rao [26].

statistical model can be updated (improved) based on observation data. Bayesian updating provides a reasonable framework for this process.

Mechanical behavior of soils shows strong nonlinearity. The nonlinearity refers to the characteristic that the response of soil to stress or strain is not directly proportional. In other words, the relationship between the applied stress and the resulting strain in soil is not a straight line, but rather it exhibits a more complex, often curved relationship. **Figure 5** shows an example of stress–strain relationship of a clayey soil. At the beginning of the curve, the relationship between stress and strain is linear. In this region, the soil deforms proportionally to the applied stress, and the deformation is usually reversible if the stress is removed. As the stress increases, the curve reaches a point where the soil begins to yield, meaning it starts to undergo plastic deformation. Beyond this point, the relationship between stress and strain is no longer linear. After the yield point, the curve enters the plastic region, where the soil continues to deform under constant or slightly increasing stress. If the stress continues to increase, the curve eventually reaches a peak. The failure point is characterized by a significant increase in strain with little or no increase in stress. Beyond the failure point, the stress decreases while the strain continues to increase.

It is clear that the stress–strain curve in **Figure 5** shows strong nonlinearlity which is a critical factor in geotechnical engineering that can affect the design and analysis of geotechnical structures such as foundations, slopes, retaining walls, and other soil-structure interaction problems. It requires sophisticated nonlinear models and analysis techniques to accurately predict soil behavior under various loading conditions.

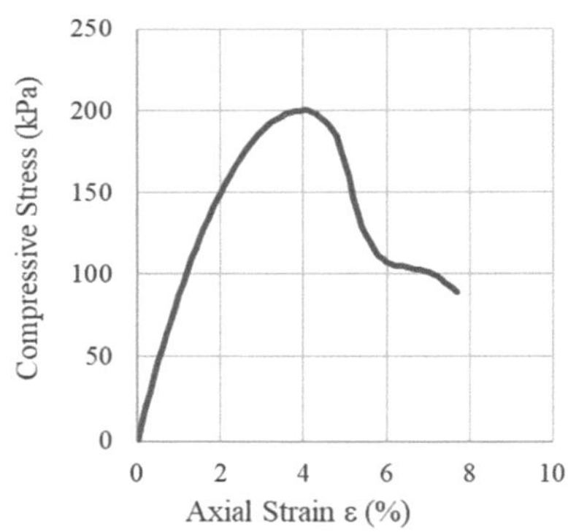

Figure 5.
Example of nonlinear behavior of soil (stress–strain curve).

Algorithm	Problem or model	Likelihood (Observation noise)	Prior probability
Kalman filter (KF)	Linear	Gaussian	Gaussian
Extended Kalman filter (EKF)	Nonlinear	Gaussian	Gaussian
Unscented Kalman filter (UKF)	Nonlinear	Gaussian	Gaussian
Ensemble Kalman filter (EnKF)	Nonlinear	Gaussian	Gaussian
Particle filter (PF)	Nonlinear	Non-Gaussian	Non-Gaussian

Table 1.
Some algorithms for sequential Bayesian filtering [7].

Table 1 summarizes several filtering algorithms. As mentioned, standard KF is designed for linear and Gaussian problems and is difficult to apply for practical geotechnical problems. Although nonlinear filtering such as EKF, UKF, and EnKF can analyze nonlinear problems, they cannot capture complex non-Gaussian probability functions. As shown in **Figure 2**, histograms of soil and rock parameters often show non-Gaussian distributions. Hence, PF is the most preferable algorithm of filtering for geotechnical applications.

3. A versatile nonlinear filter: particle filter

As discussed in Section 2, geotechnical problems are fundamentally nonlinear and non-Gaussian, hence it is natural to use PF for geotechnical problems (see **Table 1**). This section provides a thorough description of the PF.

We consider a nonlinear and non-Gaussian state space model, which is represented in the following [16]:

$$x_t = f_t(x_{t-1}) + v_t \tag{1}$$

$$y_t = h_t(x_t) + w_t \tag{2}$$

where $x_t = (x_1, x_2, \ldots, x_k)$ and $y_t = (y_1, y_2, \ldots, y_l)$ indicate the state (or parameter) vector and the observation (or data) vector, and the subscripts k and l indicate the size of the vectors. f_t and h_t are the nonlinear operator (or function) and the observation operator. Subscript t indicates the discrete time. The vectors v_t and w_t are the vectors for system and observation errors respectively. The errors follow the normal (or Gaussian) distributions with zero mean which are defined by:

$$v_t \sim N(0, Q_t) \tag{3}$$

$$w_t \sim N(0, R_t) \tag{4}$$

where Q_t and R_t indicate covariance matrices.

This chapter focuses on the mechanical behavior of geomaterials which are commonly modeled by soil–water coupled FEM. When the soil–water coupled FEM is used, Eq. (1) is given by the following equation:

$$\begin{Bmatrix} u_t \\ p_t \end{Bmatrix} = \begin{Bmatrix} u_{t-1} \\ p_{t-1} \end{Bmatrix} + \begin{bmatrix} [\mathrm{K}] & [\mathrm{K_v}]^{\mathrm{T}} \\ [\mathrm{K_v}] & -\theta\Delta t[\mathrm{K_h}] \end{bmatrix}^{-1} \begin{Bmatrix} \{\mathrm{F}\} + [\mathrm{K_v}]^{\mathrm{T}} p_t \\ \{\mathrm{Q}\} + (1-\theta)[\mathrm{K_h}]p_t \end{Bmatrix} + \begin{Bmatrix} v_t^{\mathrm{u}} \\ v_t^{\mathrm{p}} \end{Bmatrix} \tag{5}$$

where

[K]: the tangent stiffness matrix defined by constitutive models

$[\mathrm{K_v}]$: the matrix transforming the increment in nodal displacements to the volume change of each element

$[\mathrm{K_h}]$: the matrix for seepage flow in soils

{F}: a vector representing applied force

$\{u_t\}$: the displacement vector

$\{p_t\}$: the ore pressure vector

θ: coefficient $(0 < \theta < 1)$

{Q}: the vector representing increment in the volume rate of the water flow

$\{v_t^u\}$: the system error vector for $\{u_t\}$

$\{v_t^p\}$: the system error vector for $\{p_t\}$

The state (or parameter) vector x_t is defined by

$$x_t = \begin{Bmatrix} u_t \\ p_t \end{Bmatrix} \tag{6}$$

If state variables u_1, u_2, and u_3 in $\{u_t\}$ are directly observed, nonlinear function h_t can be written in matrix form as

$$H_t = \begin{bmatrix} 1 & 0 & 0 & \cdots & 0 \\ 0 & 1 & 0 & \cdots & 0 \\ 0 & 0 & 1 & \cdots & 0 \end{bmatrix} \tag{7}$$

We can describe $y_t = H_t x_t$ using Eq. (7).

The recursive formula of probability densities for prediction and filtering is defined as follows:

Prediction (time update):

$$\begin{aligned} p(x_t|y_{1:t-1}) &= \int_{-\infty}^{\infty} p(x_t, x_t|y_{1:t-1})dx_{t-1} = \int_{-\infty}^{\infty} p(x_t|x_{t-1}, y_{1:t-1})p(x_{t-1}|y_{1:t-1})dx_{t-1} \\ &= \int_{-\infty}^{\infty} p(x_t|x_{t-1})p(x_{t-1}|y_{1:t-1})dx_{t-1} \end{aligned} \tag{8}$$

Filtering (observation update):

$$p(x_t|y_{1:t}) = \frac{p(x_t, y_t|y_{1:t-1})}{p(y_t|y_{1:t-1})} = \frac{p(y_t|x_t, y_{1:t-1}) \cdot p(x_t|y_{1:t-1})}{p(y_t|y_{1:t-1})} = \frac{p(y_t|x_t) \cdot p(x_t, |y_{1:t-1})}{p(y_t|y_{1:t-1})} \tag{9}$$

where $p(x_t|x_{t-1})$ is the density of x_t given previous state vector x_{t-1}, $p(y_t|x_t)$ is the density of y_t given x_t.

In linear and Gaussian problems, the KF provides updating recursions using Eqs (8) and (9). However, integral computations in (8) and (9) are usually intractable for practical problems which usually show nonlinearity and non-Gaussianity. This is the main reason to use numerical methods such as PF in geotechnical problems.

The PF expresses PDFs with a set of Monte Carlo (MC) samples called an "ensemble" that has weights, and each MC sample is called a "particle." A filtered distribution at time $t - 1$, $p(x_{t-1}|y_{1:t-1})$, where $y_{1:t-1}$ denotes $\{y_1, y_2, \ldots, y_{t-1}\}$, is redefined with "particles" and weights:

$$p(x_{t-1}|y_{1:t-1}) \approx \frac{1}{N}\sum_{i=1}^{N} w_{t-1}^{(i)} \delta\left(x_{t-1} - x_{t-1|t-1}^{(i)}\right) \tag{10}$$

where N is the number of particles and δ is the Dirac delta function. Using this particle or MC approximation, integral computation in (8) and (9) become summations.

The particle approximation for the prediction distribution $p(x_t|y_{1:t-1})$ at time t is given by the following equation:

$$
\begin{aligned}
p(x_t|y_{1:t-1}) &= \int_{-\infty}^{\infty} p(x_t|x_{t-1})p(x_{t-1}|y_{1:t-1})dx_{t-1} \approx \sum_{i=1}^{N}\int_{-\infty}^{\infty} w_{t-1}^{(i)}\delta\left(x_{t-1}-x_{t-1|t-1}^{(i)}\right)p(x_t|x_{t-1})dx_{t-1} \\
&= \sum_{i=1}^{N} w_{t-1}^{(i)}\delta\left(x_t - f_t\left(x_{t-1|t-1}^{(i)}, v_t^{(i)}\right)\right) = \sum_{i=1}^{N} w_{t-1}^{(i)}\delta\left(x_t - x_{t|t-1}^{(i)}\right)
\end{aligned} \tag{11}
$$

where, $\{v_t^{(i)}\}_{i=1}^{N}$ is an independent and identically distributed (i.i.d.) sample set for Eq. (3). The calculation means that each particle for the prediction ensemble, $x_{t|t-1}^{(i)}$, is obtained by the direct calculation of $f_t(x_{t-1|t-1}^{(i)} | v_t^{(i)})$.

Filtering

We obtain the ensemble approximation for filtered distribution, from Eqs (9) and (11), and observation y_t by the following calculation:

$$
\begin{aligned}
p(x_t|y_{1:t}) &= \frac{p(y_t|x_t)\cdot p(x_t|y_{1:t-1})}{p(y_t|y_{1:t-1})} = \frac{p(y_t|x_t)\cdot p(x_t|y_{1:t-1})}{\int_{-\infty}^{\infty} p(y_t|x_t)\cdot p(x_t|y_{1:t-1})dx_t} \\
&= \frac{1}{\sum_j p\left(y_t|x_{t|t-1}^{(j)}\right)w_{t-1}^{(j)}}\sum_{i=1}^{N} p\left(y_t|x_{t|t-1}^{(i)}\right)w_{t-1}^{(i)}\delta\left(x_t - x_{t|t-1}^{(i)}\right) \\
&= \sum_{i=1}^{N}\tilde{w}_t^{(i)}w_{t-1}^{(i)}\delta\left(x_t - x_{t|t-1}^{(i)}\right) = \sum_{i=1}^{N} w_t^{(i)}\delta\left(x_t - x_{t|t-1}^{(i)}\right)
\end{aligned} \tag{12}
$$

where

$$
\tilde{w}_t^{(i)} = \frac{p\left(y_t|x_{t|t-1}^{(i)}\right)}{\sum_j p\left(y_t|x_{t|t-1}^{(j)}\right)w_{t-1}^{(j)}} \tag{13}
$$

If the observation system is linear, $p(y_t|x_{t/t-1}^{(i)})$ is given by Eq. (14):

$$
p\left(y_t|x_{t|t-1}^{(i)}\right) = \frac{1}{(2\pi)^{m/2}|R_t|}\exp\left[-\frac{\left(y_t - H_t\left(x_{t|t-1}^{(i)}\right)\right)^{\mathrm{T}} R_t^{-1}\left(y_t - H_t\left(x_{t|t-1}^{(i)}\right)\right)}{2}\right] \tag{14}
$$

The rest of the problem is how to sample from $p(x_t|y_{1:t})$ which is called "sampling." There are two ways of sampling methods which are Sequential Importance Resampling (SIR) and Sequential Importance Sampling (SIS). While classic PF algorithms use SIR [15, 16], SIS is known as a generalization of PF [17]. The pseudo-code of the two sampling methods is as follows:

Sequential Importance Resampling (SIR)

Step 1. Initial setup:

Generate an ensemble (set of particles) $\{x_0^{(1)}, x_0^{(2)}, \ldots, x_0^{(N)}\}$ from the initial prior probability distribution p_0. Set t = 1.

Step 2. Prediction (Forecasting):

Each ensemble member or particle $x_{t-1}^{(i)}$ evolves via nonlinear operator f_t given by Eq. (1).

Step 3. Filtering:

After measurement data y_t, is obtained, weight $w_t^{(i)}$ is calculated.

Step 4. Resampling (based on filtered distribution):

Resample new particles $\{x_t^{(1)}, x_t^{(2)}, \ldots, x_t^{(N)}\}$ N times from the set of particles or parameters $x_{t\text{-}1}^{(i)}$. The set of samples approximates filtered distribution $p(x_t|y_{1:t})$.

Set $t = t + 1$ and go back to Step 2.

Sequential Importance Sampling (SIS)

Step 1. Initial setup:

Generate a set of particles $\{x_0^{(1)}, x_0^{(2)}, \ldots, x_0^{(N)}\}$ from the initial (or prior) probability distribution.

Step 2. Prediction (Forecasting):

Each sample $x_{t\text{-}1}^{(i)}$ evolves via nonlinear operator f_t given by Eq. (1)

Step 3. Filtering:

After measurement data y_t, is obtained, weight $w_t^{(i)}$ is calculated.

Step 4. Weight (or probability) updating:

The set of weighted particles $\{x_t^{(i)}\}$ approximates filtered distribution $p(x_t|y_{1:t})$.

Set $t = t + 1$ and go back to Step 2.

Figure 6 compares the PDFs approximated using SIR and SIS.

The main question here is which sampling method, SIR or SIS, should be used for geotechnical applications. To answer the question, let us focus on the mechanical soil behavior. Soil undergoes both elastic and plastic deformation when load is applied to and reaches "critical state" under large shear deformation at constant volume and constant shear/normal stress conditions [27]. This mechanical behavior of soil can be well explained by the critical state constitutive models. In critical state soil mechanics, the mechanical state of soil is described by the stress parameters q and p and the specific volume $(1 + e)$. The history of consolidation, which causes long-term

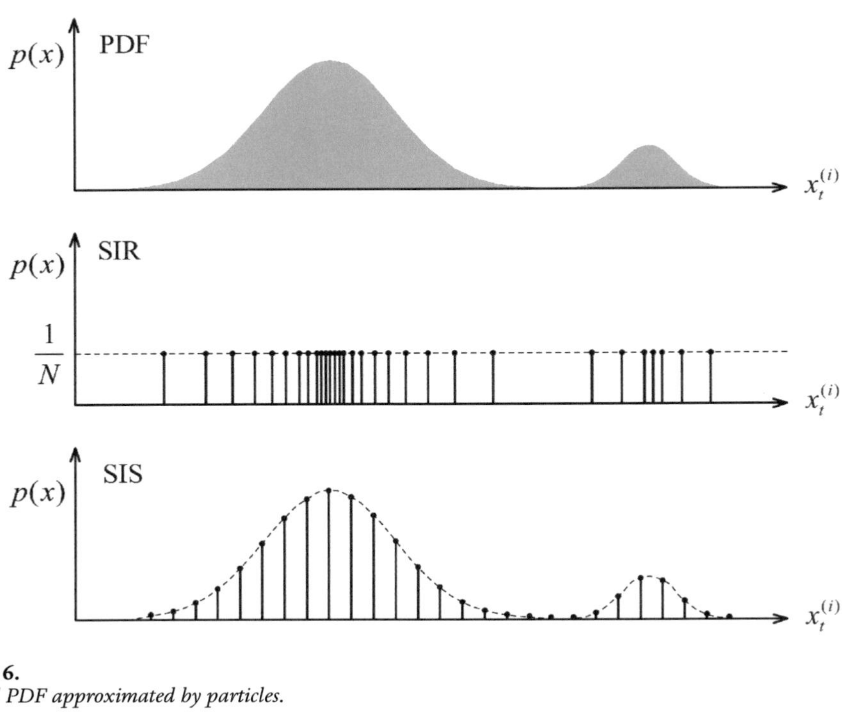

Figure 6.
Filtered PDF approximated by particles.

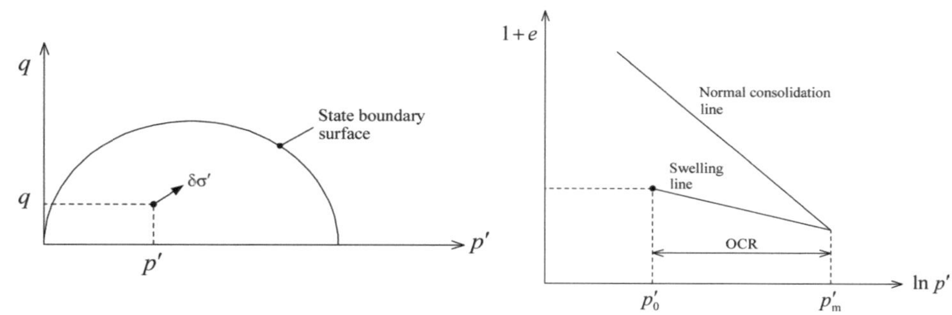

Figure 7.
Current stress and consolidation history (Modified from [28]).

settlement of the ground, is described by overconsolidation ratio p_m'/p' (p_m' is the stress at the intersection of the current swelling line with the normal consolidation line, **Figure 7**). In many theories for stress–strain behavior, the value of p_m' defines the size of the state boundary surface in **Figure 7** and it separates the elastic states inside from the elastoplastic states.

The mechanical behavior or stress–strain response of soil not only depends on the current stress state, but also on the stress history that soil has undergone [28]. The schematic on the effect of the stress history is shown in **Figure 8**. Soil samples brought

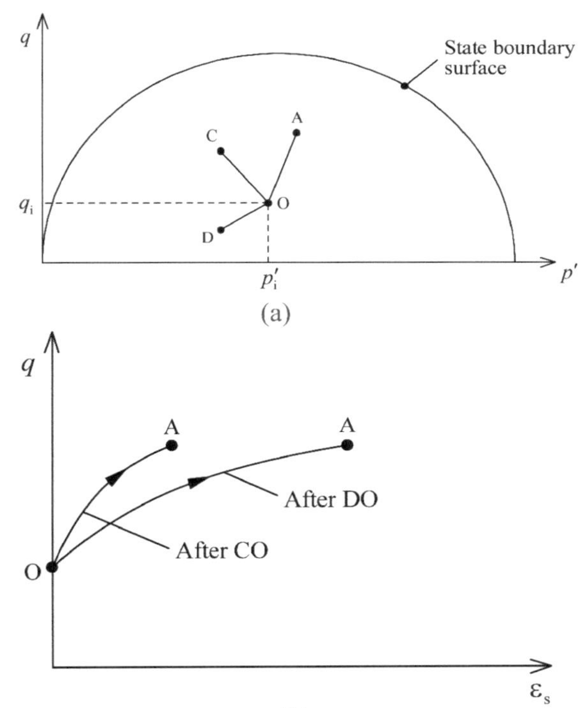

Figure 8.
Effect of recent stress history on current stiffness. (a) Stress paths of a soil, (b) stress-strain curves for the same loading path OA. (Modified from [28]).

to the same initial states of q_i and p_i at zero (O), along different paths of CO and DO, are then loaded along the same path, OA. **Figure 8b** illustrates the stress–strain curves for the same loading path OA. Soil offers different stress–strain curves depending on the stress histories undergoes before OA loading, and this is commonly observed in laboratory tests for soil samples.

Thus the stress or loading history is an essential factor for evaluating the mechanical (or stress–strain) behavior of soils. Let us refocus on the sampling methods herein. While the SIR algorithm abandons the information of stress history the soils have undergone due to resampling, the SIS keeps storing the information of stress history during the entire process of filtering. This implies that the SIS has natural ability to evaluate the mechanical behavior of soils considering stress history and it has high potential of Bayesian updating in geotechnical problems.

Selection of parameters to be identified and observed quantities [7]: The selection of parameters to be identified varies with the actual types of problems. For example, when a soft clay ground is improved by sand compaction piles or vacuum consolidation, the mechanical properties of the clay layer change after the ground improvement. Hence, the new, but unknown, properties of the improved layer, such as the hydraulic conductivity and compression index, can be a target for identification. Material constants that are difficult to measure, or parameters that significantly affect the future behavior of structures, are usually worthy of identification. Observed quantities, such as the settlement, lateral displacement, and pore water pressure, for such inverse analyses, must be sufficiently sensitive to the parameters to be identified. In practice, the measurable physical quantities are limited. Hence, we need to obtain and select the observed quantities that are relevant to the parameters to be identified.

Effective sample size [7]: When PF is used for Bayesian updating, attention must be paid to "degeneracy." The term "degeneracy" refers to the scenario whereby, after a few iterations of the particle filter algorithm, only a small number of particles have significant weight. This means that only a few particles account for most of the posterior probability mass, making the approximation less accurate. Degeneracy happens when the weights of the particles become too unequal. There is a useful metric, which is called the effective sample size [29, 30], for discussing how severe the degeneracy is. The N_{eff} is defined by the following equation:

$$N_{\text{eff}} = \frac{1}{\sum_{i=1}^{N} \left(w_k^{(i)}\right)^2} \tag{15}$$

where N is the total number of particles and ${w_k}^{(i)}$ is the weight of the ith particle at the kth filtering step. The N_{eff} takes values from 1 (only one particle has significant weight) to N (all the particles have the same weight). In practice, when the N_{eff} is less than a certain predetermined threshold for the N_{eff}, some measures (resampling or redefining the covariance matrix in likelihood) are considered. Although the "proper" threshold depends on the problems and the purpose of the Bayesian updating, $N/5$ and $N/3$ are commonly used.

4. A case study

This section presents an example that applies PF to a typical geotechnical problem: deformation of the ground under embankment loading which is predicted by a numerical model. This example was used for a student competition on machine learning in 2020 organized by Technical Committee 304 and 309 of International Society of Soil Mechanics and Geotechnical Engineering (ISSMGE) [31]. **Figure 9** shows the geometry of the cross-section of an embankment, which was constructed on layers of soft silty and clayey soils, installed with different types of instrumentation. In the figure, "M" and "VWP" indicate instruments to observe ground settlement and excess pore pressure of the ground. "HPG1" is an instrument to measure surface settlement. In this construction site, prefabricated vertical drains (PVDs) were installed to improve the ground because the ground consists of very soft clay layers. Information on the construction sequences of the embankment is shown in **Figure 10**. In the figure, four types of data measured at different locations (SP1 ~ 4) were presented. In the student contest, the unit weight of embankment fill was assumed to

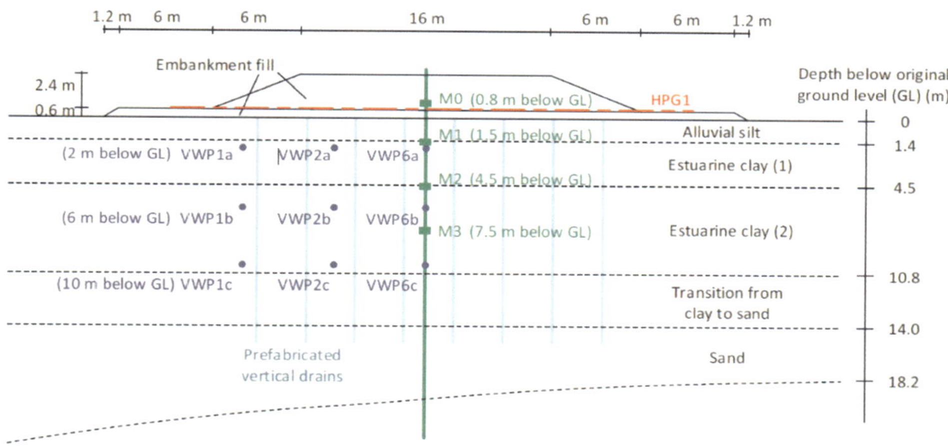

Figure 9.
Cross-section of embankment with location with instrumentation [31].

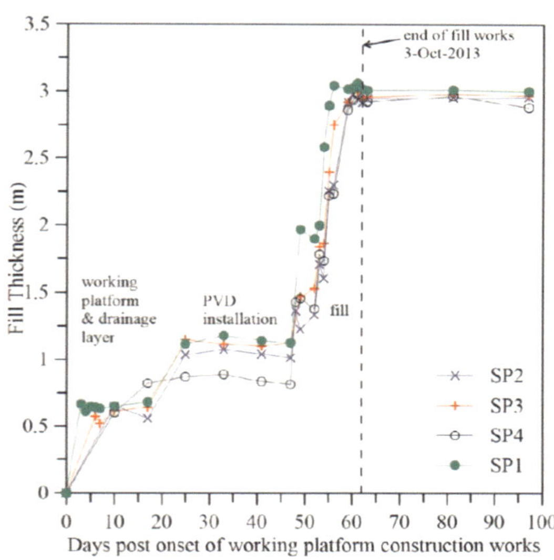

Figure 10.
Time history of embankment construction [31].

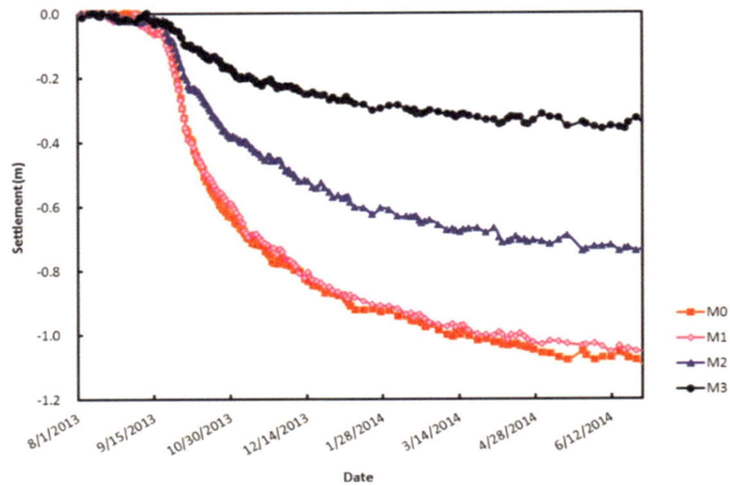

Figure 11.
Settlement data [31].

be 20 km/m^3. Deformation of the ground (settlement) was recorded at the site for around 3 years, which is shown in **Figure 11**. In the contest, only the settlement data measured for one year were presented, and the participants tried to predict the settlement after one year. The task of the contest was to develop a method to facilitate the prediction of the future settlement.

We used soil–water coupled Finite Element (FE) analysis with Cam-clay model for simulating deformation behavior of the ground. The FE mesh used for this problem is shown in **Figure 12**. We assumed horizontal displacements are zero at the edge of right and left of the model and vertical and horizontal displacements are zero at the bottom of the model. In addition, only the surface of the model ground is permeable: right, left, and bottom are assumed to be impermeable boundaries. Due to the symmetry of the embankment, only the right half was discretized for the numerical analysis. The foundation ground was divided into five layers; the first and fifth layers were assumed to be linear elastic bodies, and layers two to four were modeled as elastoplastic materials by the Cam-clay model.

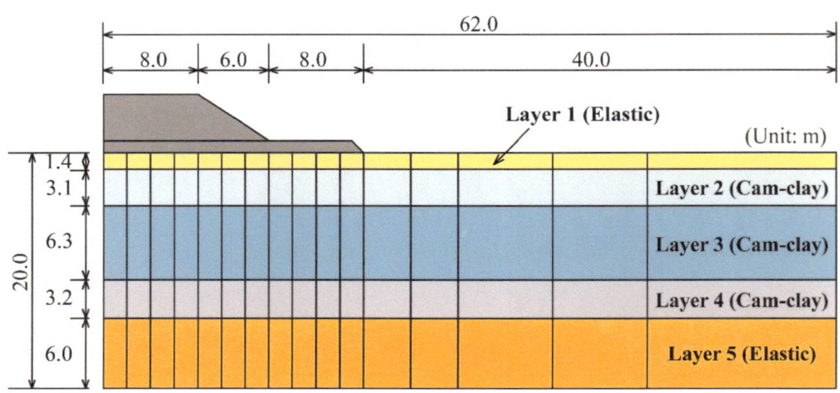

Figure 12.
FE mesh used for the simulation.

In this study, a diagonal matrix with the maximum absolute value of the observation data used for PF is employed as the covariance matrix. This assumes that level of observation noise increases with increasing output (observation data). Such an empirical setting for the covariance matrix has been used in Bayesian updating with PF in geotechnical engineering [21–23], and despite its simplicity, there is a significant track record of obtaining reasonable results. The covariance matrix is given as follows:

$$R_t = \begin{bmatrix} 1.08 & 0 & 0 & 0 \\ 0 & 1.05 & 0 & 0 \\ 0 & 0 & 0.74 & 0 \\ 0 & 0 & 0 & 0.32 \end{bmatrix} \tag{16}$$

where R_{11}, R_{22}, R_{33}, and R_{44} corresponds to the observation noise of M0, M1, M2, and M3, respectively.

In this case study, we focus only on the 2nd to 4th layers of clayey soil, which greatly influence the time-settlement behavior, and only the compression index λ and permeability coefficient k (m/d) of each layer are considered as the parameters for Bayesian updating. Therefore, the total number of geotechnical parameters to be estimated is six. Less influential parameters are assumed to be "deterministic," and they are summarized in **Table 2**.

While previous studies [21–23] mainly focused on the estimation of material parameters, this study also considers the embankment load and its loading history as unknown parameters and presents a specific example of load modeling. The reason embankment load and construction history have not been the focus previously is that nearly deterministic information can be obtained about the unit weight of the embankment and the construction process, which is considered negligible compared to the uncertainty of the foundation ground information. In this study, the uncertainty of the embankment load (f_t) at time t is modeled using the following equation:

$$f_t = \begin{cases} \alpha F_0\{(\exp(\beta t) - 1)/\exp(\beta T) - 1\} & t \leq T \\ \alpha F_0 & t > T \end{cases} \tag{17}$$

where F_0 is the load at the completion of embankment construction, f_t is the load at time t, and α and β are parameters controlling loading history of embankment. **Figure 13** represents the variation in load patterns due to differences in α when β is fixed at 1.0. Although the actual construction process shows a shape significantly different from an exponential function because of periods of rest in embankment construction, we proposed a new model based on an exponential function, which has relatively fewer parameters and is computationally easier to handle. Moreover, it is demonstrated through application examples that such a simplified modeling can

Layer	Soil type	Poisson's ratio ν	Compression index, λ Elastic modulus E (Kpa)	Swelling index, κ	Void ratio e	Critical state parameter M	Permeability k (m/d)
1, 5	Sand	0.333	128,000	-	-	-	-
2,3,4	Clay	0.348	Unknown	0.123	1.525	1.023	Unknown

Table 2.
Soil parameters assumed to be deterministic.

	Parameter	Initial distribution
Loading model	α	0.05–0.50
	β	0.8–1.2
Cam-clay model	λ	0.4–1.2
	$\log_{10}k$(m/s)	0.01–0.50

Table 3
Target parameters and their prior distributions.

function effectively in this case history. We added the parameters α and β into the unknown parameters, and the total number of parameters for Bayesian updating becomes eight. Since no prior information regarding the unknown parameters was provided, prior probabilities of the unknown parameters assumed to follow uniform distributions within the range shown in **Table 3**. During this process, several trial calculations were conducted to confirm the range within which the numerical results could explain the observational data, and then the parameter range was determined.

Generally, it is known that the required number of particles N increases in proportion to the number of estimated parameters, but as mentioned earlier, the appropriate number of particles is often determined through trial and error. In this study, calculations were performed with 50,000 particles.

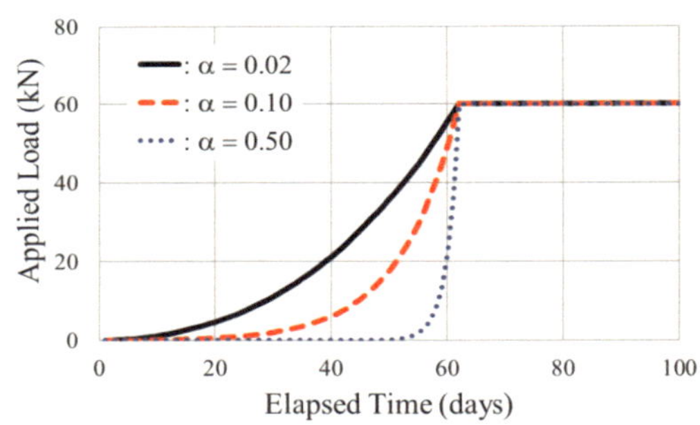

Figure 13.
Loading histories generated using Eq. (17).

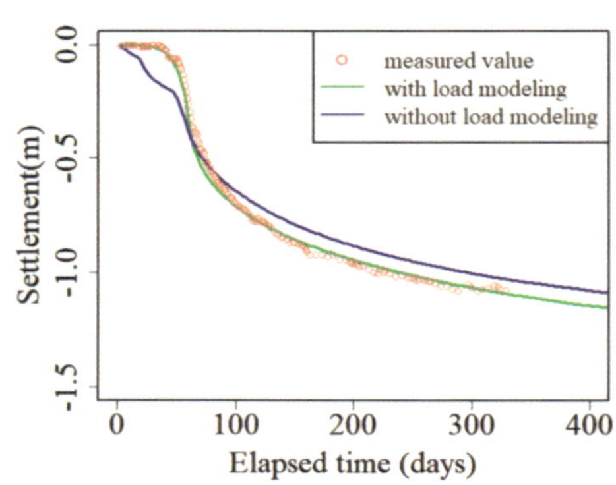

Figure 14.
Impact of the load modeling on settlement simulations.

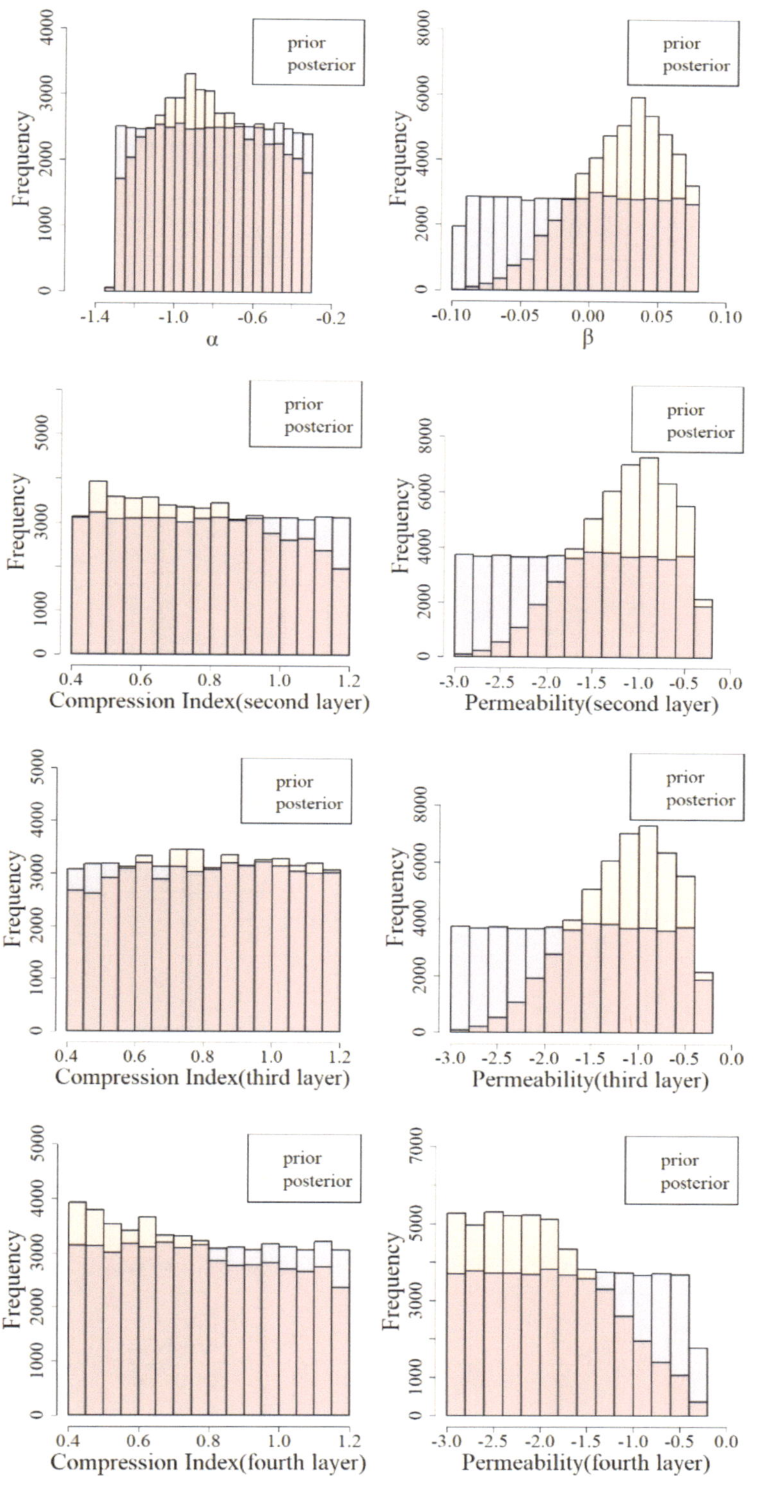

Figure 15.
Prior and posterior distributions of unknown parameters.

Figure 14 shows the impact of the presence or absence of load modeling on the Maximum A Posteriori (MAP) estimation of settlement. As evident from the figure, the load modeling enables the numerical analysis to successfully reproduce the

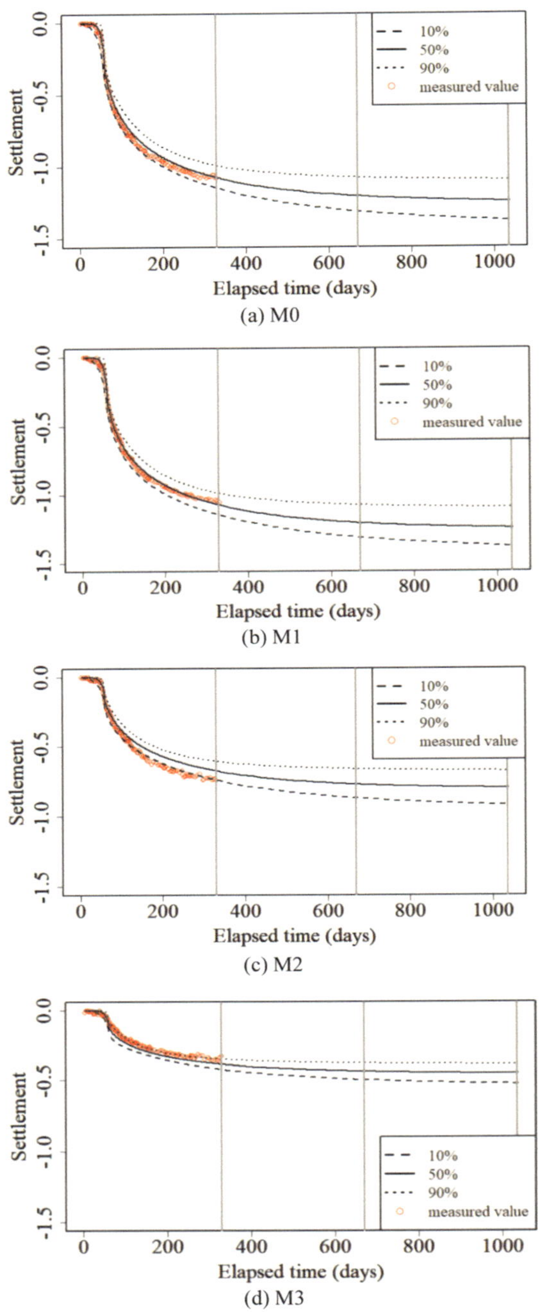

Figure 16.
Posterior distributions of settlement measured at different locations.

observation data. On the other hand, without the embankment load modeled, a significant discrepancy between measurement and prediction is observed during the embankment loading period. This can be attributed to 'modeling errors' in the simulation model representing soil mechanical behavior. While altering the simulation model (Cam-clay model) may resolve this issue, the use of load modeling proposed in this paper leads to successful simulation of actual ground.

Figure 15 shows the prior and posterior distributions for the eight unknown parameters. These histograms are drawn using 50,000 particles. Representing parameters as probability distributions allows for discussion of the uncertainty and reliability of the estimates, as well as the sensitivity of the parameters to the amount of settlement. For example, focusing on the compression index λ of the third layer, the posterior distribution is flat, showing little difference from the prior distribution. This indicates that this parameter has low sensitivity to settlement and that any value it takes does not significantly the result. On the other hand, the posterior distribution of the permeability coefficient k of the third layer is updated with a clear peak, indicating high sensitivity to settlement. In terms of sensitivity trends, the parameters of the load model (α and β) and the permeability coefficients k are highly sensitive, while the compression indices λ are less so. This can be attributed to the compression indices λ being in a trade-off relationship with the load intensity parameters (especially α) and due to the presence of multiple layers. The compression index λ is a parameter that defines the amount of settlement due to loading, and in the case of a single layer of clay ground, there are two possible combinations of parameters that can explain the amount of settlement. In other words, there are countless combinations of parameters that can explain the observational data, corresponding to the 'ill-posed problem' in the context of inverse problems, a challenge commonly encountered in analyses of inverse problems not only in geotechnical engineering but also in civil engineering.

Figure 16 shows a comparison between the settlement observed at M1 and the posterior distribution of settlement estimated by PF. These results consider load modeling, and hereafter, only results involved with load modeling are presented. The figure includes the 10%, 50%, and 90% percentile values. The percentile value represents the threshold below which 10% of the PF posterior distribution samples fall when ordered from smallest to largest. This allows for discussion of the accuracy of the prediction. Due to the settings in the simulations (covariance matrix $\mathbf{\Sigma}$), the results of M0 and M1 have a significant impact, and the predictions match these two results. As a result, in M2 and M3, the measurement data do not fall within the 10% – 90% range, leading to an underestimation of settlement in M2 and an overestimation in M3. In this study, since the settlement at the ground surface is the most important for embankment structures, Bayesian updating focused on the prediction accuracy of surface settlement was conducted. However, if one wishes to focus on predictions for M2 or M3, adjustments can be made in the covariance matrix. By reducing the variance of the focused observation point, its influence becomes more significant in the Bayesian updating.

Figure 17 shows the posterior distribution of the settlement for M0–M3 as of June 1, 2016. The figure also displays each percentile value. The posterior distribution shows a nearly symmetrical Gaussian distribution shape, and irrespective of the measurement point, the 50% percentile value almost coincides with the mean of the distribution. It has been reported in previous literature that the posterior distribution being unimodal and symmetrical is common. This suggests the possibility of applying

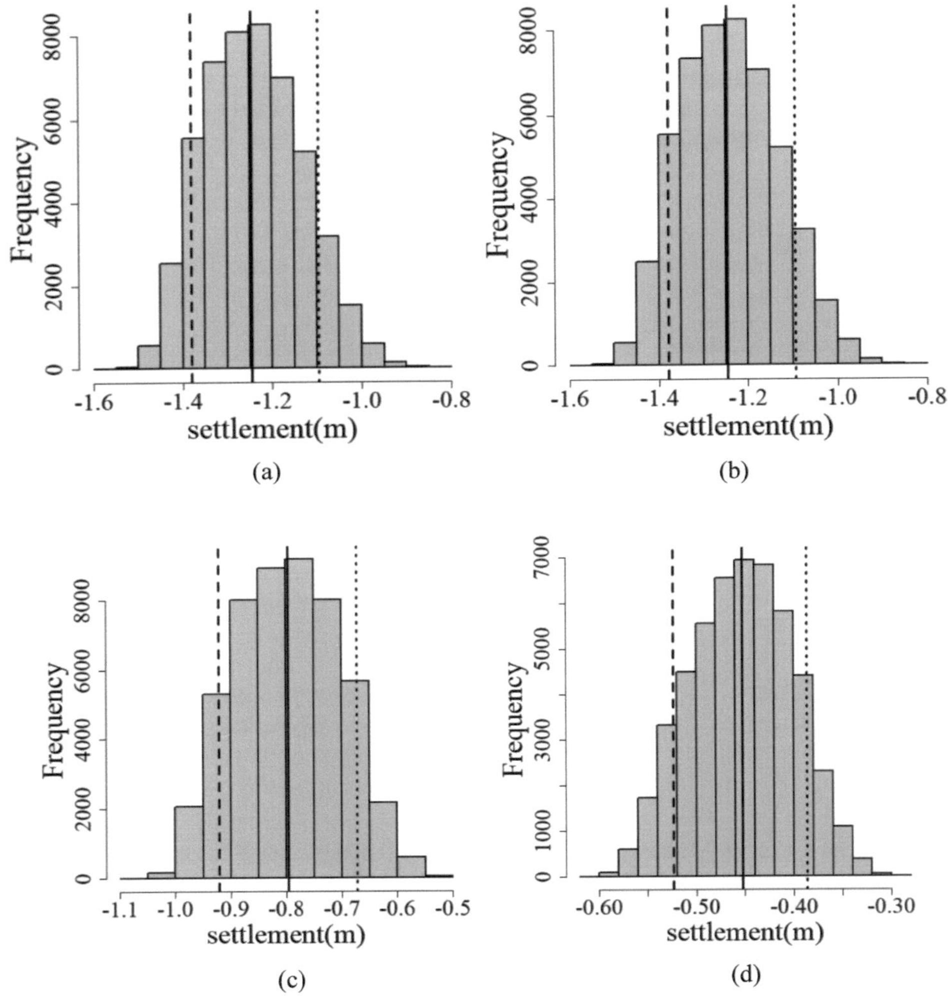

Figure 17.
Posterior distributions of settlement at June 1, 2016.

more efficient algorithms than PF in ground analysis, which is a subject for future research.

5. Conclusion

This chapter presented the theoretical and the practical effectiveness of the PF for geotechnical problems through a case history. First, some essential characteristics of geotechnical problems such as heterogeneity of the ground and variability of soil/rock parameters are explained to emphasize the motivation for using probabilistic methods in geotechnical engineering. Second, we have outlined the fundamentals PF by comparing two different sampling algorithms (SIR and SIS) and discussed which algorithm is more preferable in geotechnical problems from an aspect of geomechanics:

DOI: http://dx.doi.org/10.5772/intechopen.1005562

stress history dependency of soils. Then, PF was applied to a real case history on embankment construction on clay ground to discuss the effectiveness in parameter (or probability density) identification and prediction in geotechnical engineering. The main conclusions of this study can be summarized as follows:

1. Probability estimation or uncertainty quantification plays a crucial role in geotechnical engineering because several different types of uncertainties (e.g., heterogeneity of the ground and variability of soil parameters) can be included in geotechnical design and construction process. Hence it is natural to use parameter identification methods that can provide posterior probability distributions of estimate such as KF and nonlinear KF in geotechnical engineering.

2. Unlike linear KF, the PF is applicable to nonlinear and non-Gaussian problems. Geotechnical problems usually show nonlinearity and non-Gaussianty, and the PF is preferable to apply in geotechnical engineering compared to standard KF.

3. Two types of sampling methods, SIR and SIS, are used in PF. Although SIR is commonly used in many different research fields, SIS is preferable to use for geotechnical problems because SIS can consider stress history soils undergo, which greatly impacts on mechanical behavior of soils.

4. PF can accurately estimate the posterior PDFs of geotechnical parameters in soil–water coupled FEM. In addition, PF can also be applied to identification of boundary condition, loading history in this study, for numerical simulations. The simulation with identified parameters predicts the future behavior of the ground with high accuracy: these results show that the PF has high potential for more general parameter identification and prediction problems in geotechnical engineering.

Author details

Akira Murakami[1*], Takayuki Shuku[2] and Kazunori Fujisawa[3]

1 Graduate School of Advanced Integrated Studies in Human Survivability, Kyoto University, Kyoto, Japan

2 Department of Urban Engineering, Tokyo City University, Tokyo, Japan

3 Graduate School of Agriculture, Kyoto University, Kyoto, Japan

*Address all correspondence to: akiram@kais.kyoto-u.ac.jp

References

[1] Phoon KK, Shuku T, Ching J. Uncertainty, Modeling, and Decision Making in Geotechnics. Boca Raton, FL, USA: CRC Press; 2023. 520 p. DOI: 10.1201/9781003333586

[2] Kalman RE. A new approach to linear filtering and prediction problems. Journal of Basic Engineering;**82**:35-45

[3] Murakami A, Hasegawa T. Observational prediction of settlement using Kalman filter theory. Numerical Methods in Geomechanics. 1985;**3**: 1637-1643

[4] Murakami A, Hasegawa T. Back analysis by Kalman Filter-finite elements and a determination of optimal observed points location. Proceedings of the JSCE. 1987;**388**:227-235 (in Japanese)

[5] Murakami A, Hasegawa T. Back analysis by Kalman Filter-finite elements and optimal location of observed points. Numerical Methods in Geomechanics. 1988;**3**:2051-2058

[6] Murakami A. Studies on the Application of Kalman Filtering to Some Geomechanical Problems Related to Safety Assessment. Ph.D. dissertation. Kyoto University; 1991

[7] Murakami A, Fujisawa K, Shuku T. Developments of inverse analysis by Kalman filters and Bayesian methods applied to geotechnical engineering. Proceedings of the Japan Academy Series B. 2023;**99**(9):352-388

[8] Stefano M, Corigliano A. Impact induced composite delamination: state and parameter identification via joint and dual extended Kalman filters. Computer Methods in Applied Mechanics and Engineering. 2005;**194**: 5242-5272

[9] Yang C, Wu Y, Hon T, Feng XT. Application of extended Kalman filter to back analysis of the natural stress state accounting for measuring uncertainties. International Journal for Numerical and Analytical Methods in Geomechanics. 2011;**35**:694-712

[10] Nguyen LT, Datcheva M, Nestorovic T. Identification of a fault zone ahead of the tunnel excavation face using the extended Kalman filter. Mechanics Research Communications. 2013;**53**:47-52

[11] Wang Y, Jin H, Ouyang L. Real-time prediction of seepage field during tunnel excavation. Applied Mechanics and Materials. 2013;**274**:11-16

[12] McElhoe BA. An assessment of the navigation and course corrections for a manned Flyby of Mars or Venus. IEEE Transactions on Aerospace and Electronic Systems. 1966;**2**(4):613-623

[13] Julier SJ, Uhlmann JK. A new extension of the Kalman filter to nonlinear systems. In: Proc. SPIE, Signal Processing Sensor Fusion and Target Recognition IV. 1997. pp. 182-193

[14] Evensen G. Sequential data assimilation with a non-linear quasi-geostropic model using Monte Carlo methods to forecast error statistics. Journal of Geophysical Research. 1994; **99**:10143-10621

[15] Gordon NJ, Salmond DJ, Smith AFM. Novel approach to nonlinear/non-Gaussian Bayesian state estimation. IEE Proceedings-F. 1993;**140**(2):107-113

[16] Kitagawa G. Monte Carlo filter and smoother for non-Gaussian nonlinear state space mode. Journal of

Computational and Graphical Statistics. 1996;**5**(1):1-25

[17] Doucet A, Godsill S, Andrieu C. On sequential Monte Carlo sampling methods for Bayesian filtering. Statistics and Computing. 2000;**10**:197-208

[18] Chen Y, Zhang D. Data assimilation for transient flow in geologic formations via ensemble Kalman filter. Advances in Water Resources. 2006;**29**:1107-1122

[19] Hommels A, Murakami A, Nishimura S. A comparison of the ensemble Kalman filter with the unscented Kalman filter: application to the construction of a road embankment. GEO International. 2009;**13**(1):52-54

[20] Liu K, Vardon PJ, Hicks MA. Sequential reduction of slope stability uncertainty based on temporal hydraulic measurements via the ensemble Kalman filter. Computers and Geotechnics. 2018; **95**:147-161

[21] Shuku T, Murakami A, Nishimura S, Fujisawa K, Nakamura K. Parameter identification for Cam-Clay model in partial loading model tests using the particle filter. Soils and Foundations. 2012;**52**(2):279-298

[22] Murakami A, Shuku T, Nishimura S, Fujisawa K, Nakamura K. Data assimilation using the particle filter for identifying the elasto-plastic material properties of geomaterials. International Journal for Numerical and Analytical Methods in Geomechanics. 2013;**37**: 1642-1669

[23] Shibata T, Shuku T, Murakami A, Nishimura S, Fujisawa K, Hasegawa N, et al. Prediction of long-term settlement and evaluation of pore water pressure using particle filter. Soils and Foundations. 2019;**59**:67-83

[24] Shuku T, Phoon KK. Comparison of data-driven site characterization methods through benchmarking: Methodological and application aspects. ASCE-ASME Journal of Risk and Uncertainty in Engineering Systems, Part A: Civil Engineering. 2023;**9**(2):1-18. DOI: 10.1061/AJRUA6.RUENG-977

[25] Christian JT, Ladd CC, Beacher GB. Reliability applied to slope stability analysis. Journal of Geotechnical Engineering;**120**(2):2180-2207

[26] Rao CR. Statistics and Truth. 2nd ed. Singapore: World Scientific Publishing; 1997. 212 p. DOI: 10.1142/3454

[27] Schofield AN, Wroth CP. Critical State Soil Mechanics. New York, United States: McGraw-Hill; 1968

[28] Atkinson JH, Richardson D, Stallebrass SE. Effect of recent stress history on the stiffness of overconsolidated soil. Géotechnique. 1990;**40**(4):531-540

[29] Liu JS. Monte Carlo Strategies in Scientific Computing. New York: Springer; 2001 364 p

[30] Kitagawa G, Takemura A. Mathematical and Computational Statistical Science. In: Statistical Science in the 21st Century, III. Meguro-ku, Tokyo, Japan: The University of Tokyo Press; 2008. pp. 297-298 (in Japanese)

[31] Website of TC304, ISSMGE. TC304-TC309 student contest of machine learning (APSSRA2020, Tokyo, Japan). 2020. Available from: http://140.112.12.21/issmge/tc304.htm?=10

Chapter 3

Data Sensor Fusion for Surveillance Applications: Evaluation of Extended Kalman Filter vs. Unscented Kalman Filter

Oscar M. Sogamoso, Eduardo A. Fernández and Marco J. Suarez

Abstract

This chapter introduces the foundations of the Extended Kalman Filter (EKF) and Unscented Kalman Filter (UKF) for data sensor fusion applications in surveillance applications. After explaining how to model the drone under the constant turn rate and velocity (CTRV) dynamics, the EKF and UKF techniques for Lidar and Radar data sensor fusion are applied to enable better 3D object detection and reconstruction from point cloud data is evaluated under a performance comparison. Through root mean square error (RMSE) and mean square error (MSE) the filters are designed to meet the specifications of the model based on the position of the gathered points (X and Y axis) in the space state model. By degrading the measurements with Gaussian and Rayleigh noise, the results evaluate and compare both techniques to show the tolerance and performance in terms of stability, scalability, and application to drone monitoring in surveillance activities.

Keywords: extended Kalman filter, unscented Kalman filter, data sensor fusion, Lidar, radar, cloud point, CTRV model

1. Introduction

Intelligent autonomous systems have gained great importance today, often supported by artificial intelligence, assisting in daily tasks by reducing margins of error associated with human factors. Specifically in the field of autonomous driving, new trends in automotive manufacturing focus on the implementation of self-sufficient techniques for driver assistance, saving lives, and reducing driver fatalities on highways. It is estimated that 90% of accidents are due to human error, and 40% of these errors are caused by driver distraction. One of the first prospective studies in this regard was carried out in 2011 by the GDV, the German Insurance Association, which concluded that systems that detect vehicles, as well as pedestrians, can reduce the number of collisions by 30.7%, and if they also detect cyclists, they could avoid up to 45.4% of accidents involving these users [1].

Deep Learning and Machine Learning are effective tools for the creation of autonomous driving systems, which employ Lidar sensors, infrared radar, and cameras to generate spatial awareness for correct decision-making. However, the inherent uncertainty of the sensors can lead to failures in taking data from the environment, thus reducing the safety and robustness of the system, which in the first instance was designed to avoid accidents. To counteract this phenomenon, sensor data fusion is employed, which achieves a reduction in the uncertainty in the data obtained from the environment by reducing the gap between measurements and actual data [2, 3]. Data fusion can occur in sensor arrays that use similar physical phenomena for measurement, such as sensor arrays that are heterogeneous to each other.

The EKF is one of the widely used methods for sensor data fusion [4], this chapter will show the design of Lidar and Radar sensor fusion starting from the CTRV model [4, 5] and how it can be taken to a three-dimensional model for a drone-type flying device. Performance comparison will also be made against the UKY model.

2. CTRV model for EKF

To extend the CTRV model to three dimensions, the following assumptions were taken into account:

- Automobiles on highways do not present tangential movements between them, since the common traffic on highways maintains trajectories with a regular tendency.
- The projection of the drone position will be examined as a function of the angular velocity of the drone and the angular velocity of the tracked object.

Figure 1 shows the dynamics between two cars on a highway, where ρ is the magnitude of the distance between the car with the sensor data fusion and the target, θ is the reference angle between the two, and ω is the angular velocity of the target.

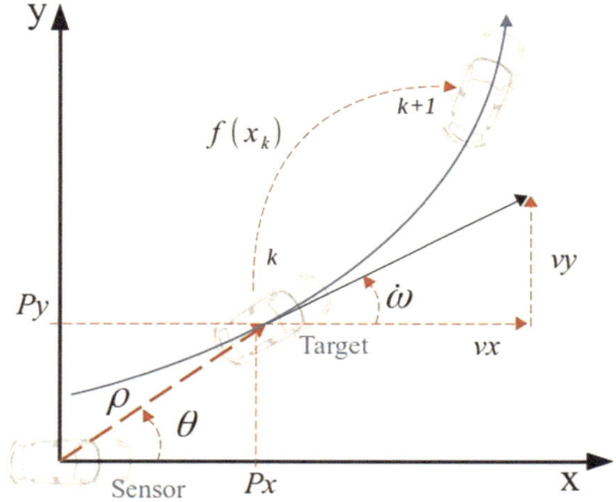

Figure 1.
CTRV model for dynamics between two automobiles.

The CTRV model between the UAV system - moving target for the three-dimensional case, determines the projection of the position of the target $x_i + 1$ on the axis starting from the values of the angular frequency w and the angle θ, [4, 5]. The commonly used sensors provide x and y positions concerning the measurement device in the case of Lidar and distance and angle formed with the measurement device in the case of Radar. Therefore, these variables are contemplated in the state variables x, y, and θ.

$$\overline{x} = [x, y, v, \theta, w, w_d] \tag{1}$$

Therefore, the equations that model the dynamics in tracking the final position of the object, as a function of the angular velocities of the sensor drone and the velocity of the target have the following form:

$$\overline{x}_{i+1} = x_i + [v_{object} - v_{drone}] \cdot \Delta T \tag{2}$$

$$\overline{y}_{i+1} = y_i + [v_{object} - v_{drone}] \cdot \Delta T \tag{3}$$

The state variables are the frontal velocity, the theta angle, the target angular velocity, and the angular velocity of the UAV.

$$\begin{aligned} v &= w \cdot \Delta T \\ \theta &= 0 \\ w &= 0 \\ w_d &= 0 \end{aligned} \tag{4}$$

Because the data sensor fusion operation is bidimensional, the CTRV model does not include motion in the position about the z-axis in its state variables. To maintain a bi-dimensional analysis, the UAV velocity is projected as:

$$v_x = V \cos \phi \tag{5}$$

$$v_y = V \sin \phi \tag{6}$$

In this way ϕ represents the elevation of the UAV concerning the sensed target, this angle allows to projection of the velocities of the drone in the xz plane and determines the $x + \Delta x$ or $y + \Delta y$, as shown in **Figure 2**, for the position prediction. The current technology provides IMU (inertial measurement units) and its evolution to AHRS (Attitude and Heading Reference Systems) shown in **Figure 3**, which internally incorporates a fusion of Gyroscope Accelerometer and magnetometer to achieve a georeferencing and is able to project the velocities of the drone to an XY plane to maintain the philosophy of the CTRV model.

To simplify the physical model for the EKF, we propose a data acquisition method in which the UAV only uses pitch (rotation on the lateral Y axis) and yaw (rotation on the vertical Z-axis) movements, and their projection in a three-dimensional coordinate system [6].

$$v_d = \begin{bmatrix} V_{dx} \\ V_{dy} \\ V_{dz} \\ W_{dz} \end{bmatrix} = \begin{bmatrix} V_0 \cos\phi \cos\theta \\ V_0 \cos\phi \sin\theta \\ V_0 \sin\phi \\ 0 \end{bmatrix} \tag{7}$$

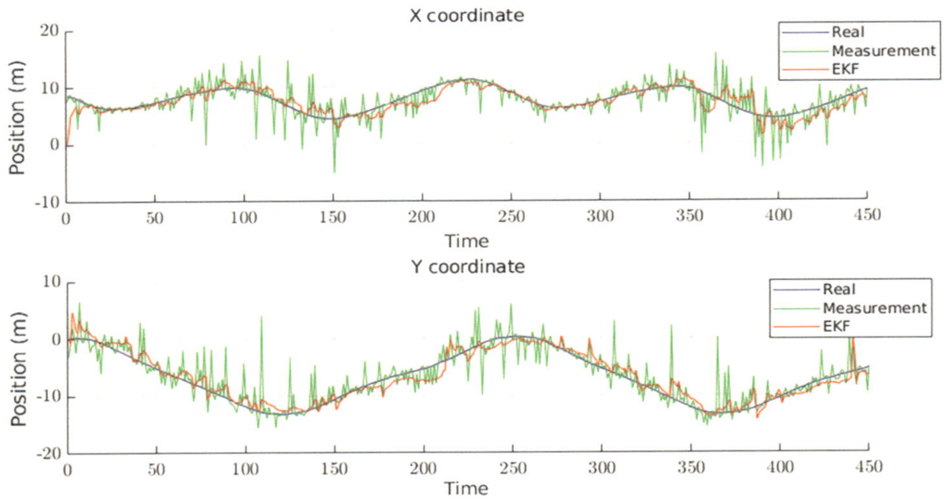

Figure 2.
EKF performance for x and y position estimation.

Figure 3.
AHRS device.

The angular velocity of the target is possible to measure from the radar sensor, so the change of position of the target is also a function of the angular velocity as $\Delta T \cdot w$. The velocity equations are obtained from x_i and y_i, which corresponds to its first derivative, such that $\dot{x}_{i+1}$ and $\dot{y}_{i+1}$ are expressed as:

$$\begin{aligned} \dot{x}_{i+1} &= \frac{v}{w}[sin\,(\Delta T \cdot w + \theta) - sin\theta] - v_{dx}\cos\theta\cos\phi \\ \dot{y}_{i+1} &= \frac{v}{w}[\cos\theta - \cos(\Delta T \cdot w + \theta)] - v_{dy}\cos\theta\sin\phi \\ a &= \dot{w} \\ \dot{\theta} &= 0 \\ \dot{w} &= 0 \\ \dot{w}_d &= 0 \end{aligned} \tag{8}$$

When the target has an initial angular velocity w = 0, the expressions change to [5]:

$$\begin{aligned} \overline{x}_{i+1} &= x_i + v\cos\theta \cdot \Delta T - v_{dx}\cos\theta\cos\phi \cdot \Delta T \\ \overline{y}_{i+1} &= y_i + v sin\theta \cdot \Delta T - v_{dx}\cos\theta sin\phi \cdot \Delta T \\ v &= 0 \\ \theta &= 0 \\ w &= 0 \\ w_d &= 0 \end{aligned} \tag{9}$$

For the spatial projection of the data obtained from the fusion of Lidar and Radar sensors, a system with spherical coordinates is used.

The EKF allows us to reduce the uncertainty around the variables measured from the noisy sensor, by executing the following steps: (i) Projection of the system dynamics as a function of the state variables previously calculated, the error covariance matrix associated with the system, and the covariance matrix associated with the noise of the observations. (ii) The calculation of the Kalman gain allows to minimize the error of the current prediction with respect to it. (iii) Finally, the algorithm update of the measurement matrix and the error covariance matrix associated with the system. The representation from the selected state variables corresponds to the following equation:

$$\overline{x} = f(x) + w \tag{10}$$

Where $\overline{x}$ is the vector of states of the system and $f(x)$ is a nonlinear function of those states. This model in state space is based on the dynamics of the system and allows to project of an estimation of the states and outputs a priori from the previous states. The reduction of the error between the values measured by the sensors and the values estimated by the filter is performed in an iterative way [7], mainly involving the projection of the state variables: The Kalman gain, the values measured by the sensors, the error covariance matrix associated to the system, and the covariance matrix associated to the noise of the observations.

$$M_k = \Phi_k P_k \Phi_k^T + Q_k \tag{11}$$

The Kalman gain, the update of the measurements, and of the state variables, is defined according to the following equations:

$$K_k = M_k H^T \left(H M_k H^T + R_k\right)^{-1} \tag{12}$$

$$P_k = (I - K_k H) M_k \tag{13}$$

$$\overline{z}_k = H\overline{x} + V_k \tag{14}$$

$$\overline{x}_k = \Phi_k \overline{x}_{k-1} + K_k (z_k - H\Phi_k \overline{x}_{k-1}) \tag{15}$$

The fundamental matrix for the EKF is defined as:

$$F = \left.\frac{\delta f(x)}{\delta x}\right|_{x=\hat{x}} \tag{16}$$

The Jacobian corresponds to the first partial derivative of each state variable, it performs a linearization of the F function as can be seen next:

$$J = \frac{\delta \overline{x}}{\delta(x, y, v, \theta, w, w_d)} \tag{17}$$

Therefore, the discrete fundamental matrix corresponds to:

$$\Phi_k = I + F \cdot \Delta T \tag{18}$$

$$\Phi_k = \begin{bmatrix} 1 & 0 & \frac{1}{w}(-sin\theta + sin\ \kappa) & \frac{v}{w}(-cos\theta + cos\ \kappa) & \frac{\Delta Tv}{w}\left[cos\ \kappa - \frac{v}{w^2}(-sin\theta + sin\ \kappa)\right] & -v_d sin\theta cos\Phi \\ 0 & 1 & \frac{1}{w}(cos\theta - cos\ \kappa) & \frac{v}{w}(-sin\theta + sin\ \kappa) & \frac{\Delta Tv}{w}\left[sin\ \kappa - \frac{v}{w^2}(cos\theta - cos\ \kappa)\right] & v_d cos\theta sin\Phi \\ 0 & 0 & 1 & \Delta T & 0 & 0 \\ 0 & 0 & 0 & 1 & 0 & 0 \\ 0 & 0 & 0 & 0 & 1 & 0 \\ 0 & 0 & 0 & 0 & 0 & 1 \end{bmatrix} \tag{19}$$

In Eq. 19, $\kappa = \Delta T + \theta$.

In case of angular velocity of the target is zero, the fundamental matrix changes to:

$$\Phi_k = \begin{bmatrix} 1 & 0 & cos\theta \cdot \Delta T & -v \cdot sin\theta \cdot \Delta T & 0 & -v_d sin\theta cos\phi \\ 0 & 1 & sin\theta \cdot \Delta T & v \cdot \cos\theta \cdot \Delta T & 0 & v_d cos\theta sin\phi \\ 0 & 0 & 1 & 0 & 0 & 0 \\ 0 & 0 & 0 & 1 & 0 & 0 \\ 0 & 0 & 0 & 0 & 1 & 0 \\ 0 & 0 & 0 & 0 & 0 & 1 \end{bmatrix} \tag{20}$$

The matrix associated with the system noise was calculated from the discrete output matrix. For the proposed model the angular acceleration of the target has been taken into account; as well as the angular velocity and acceleration of the UAV μ_a, μ_w, μ_{wd}. The output matrix for the EKF is presented below.

$$G_k \mu = \int_0^{Ts} \Phi_k(\tau) \cdot G \cdot d(\tau) \tag{21}$$

Therefore, replacing the solution in matrix form is:

$$G_k \mu = \begin{bmatrix} \frac{\mu_a}{2}\Delta T \cos\theta & 0 & -\mu_{wd}\Delta T \cos\theta \cos\phi \\ \frac{\mu_a}{2}\Delta T^2 \cos\theta & 0 & -\mu_{wd}\Delta T \cos\theta \cos\phi \\ \Delta T & 0 & 0 \\ 0 & 0 & 0 \\ 0 & \Delta T & 0 \\ 0 & 0 & \Delta T \end{bmatrix} \cdot \begin{bmatrix} \mu_a \\ \mu_w \\ \mu_{wd} \end{bmatrix} \tag{22}$$

The noise matrix from the output matrix is calculated through the following expression:

DOI: http://dx.doi.org/10.5772/intechopen.1005229

$$Q_k = G_k \cdot E\left[\mu \cdot \mu^T\right] \cdot G_k^T \tag{23}$$

With E being the expectation or mean value of the variance in the state variables magnitude, angular velocity of the target, and the angular velocity of the drone.

$$E\left[\mu \cdot \mu^T\right] = \begin{bmatrix} \sigma_a^2 & 0 & 0 \\ 0 & \sigma_w^2 & 0 \\ 0 & 0 & \sigma_{wd}^2 \end{bmatrix} \tag{24}$$

Where σ_a^2 is the standard deviation for the magnitude of the distance detected by the radar, σ_w^2 is the standard deviation in the angular velocity extracted from the radar, σ_{wd}^2 and is the angular velocity measured by the IMU inside the drone:

The covariance matrix associated with the noise of the observations Q_k is represented by:

$$Q_k = \begin{bmatrix} Q_{K_{11}} & Q_{K_{12}} & \frac{\Delta T^3}{2}{\sigma_a}^2 sin\,\theta\cos\theta & 0 & 0 & -(\Delta T \cdot \sigma_{wd})^2 \cos\theta\cos\phi \\ Q_{K_{21}} & Q_{K_{22}} & \frac{\Delta T^3}{2}{\sigma_a}^2 sin\,\theta & 0 & 0 & -(\Delta T \cdot \sigma_{wd})^2 \cos\theta\sin\phi \\ \frac{\Delta T^3}{2}{\sigma_a}^2\cos\theta & \frac{\Delta T^3}{2}{\sigma_a}^2 sin\,\theta & \Delta T^2\sigma_a^2 & 0 & 0 & 0 \\ 0 & 0 & 0 & \left(\frac{\Delta T^2}{2}\sigma_{\dot{w}}\right)^2 & \frac{\Delta T^3}{2}\sigma_{\dot{w}} & 0 \\ 0 & 0 & 0 & \frac{\Delta T^3}{2}\sigma_{\dot{w}} & \Delta T^2{\sigma^2}_{\dot{w}} & 0 \\ -(\Delta T \cdot \sigma_{wd})^2\cos\theta\cos\phi & -(\Delta T \cdot \sigma_{wd})^2\cos\theta\sin\phi & 0 & 0 & 0 & (\Delta T \cdot \sigma_{wd})^2 \end{bmatrix} \tag{25}$$

Where:

$$Q_{K_{11}} = \left(\frac{\Delta T^2}{2}\sigma_a\cos\theta\right)^2 + (\Delta T \cdot \sigma_{wd}\cos\theta\cos\phi)^2 \tag{26}$$

$$Q_{K_{12}} = \left(\frac{\Delta T^2}{2}\sigma_a\right)^2 sin\theta\cos\theta + (\Delta T \cdot \sigma_{wd}\cos\theta)^2\cos\phi\sin\phi \tag{27}$$

$$Q_{K_{21}} = \left(\frac{\Delta T^2}{2}\sigma_a\right)^2 sin\theta\cos\theta + (\Delta T \cdot \sigma_{wd}\cos\theta)^2\cos\phi\sin\phi \tag{28}$$

$$Q_{K_{22}} = \left(\frac{\Delta T^2}{2}\sigma_a\right)^2 sin\theta\cos\theta + (\Delta T \cdot \sigma_{wd}\cos\theta\sin\phi)^2 \tag{29}$$

Regards the variables provided by the sensors, often in a sensor data fusion the data formats and protocols used by the manufacturers are different. In the LiDAR case, the distance estimation for autonomous driving applications is delivered in rectangular x and y coordinates, which for the purposes of this chapter are part of the state vector. In a fusion of LiDAR and Radar sensors alternated the measurements delivered by each of the sensors through the matrix H because there are six state variables to operate only the x and y variables maintaining the conformable product between $\overline{x}$ and HL, we have that:

$$H_L = \begin{bmatrix} 1 & 0 & 0 & 0 & 0 & 0 \\ 0 & 1 & 0 & 0 & 0 & 0 \end{bmatrix} \tag{30}$$

For the Radar, the measurement matrix changes to:

$$H_R = \begin{bmatrix} 1 & 0 & 0 & 0 & 0 & 0 \\ 0 & 1 & 0 & 0 & 0 & 0 \\ 0 & 0 & 1 & 0 & 0 & 0 \\ 0 & 0 & 0 & 1 & 0 & 0 \\ 0 & 0 & 0 & 0 & 1 & 0 \end{bmatrix} \tag{31}$$

The measurement error covariance matrix of the LiDAR sensor, obtained from the statistical analysis of the data set obtained from this sensor is as follows:

$$R_{k-L} = \begin{bmatrix} 0.0222 & 0 \\ 0 & 0.0222 \end{bmatrix} \tag{32}$$

R_{k-L} is obtained from the variance in the LiDAR data set. Likewise, the measurement error covariance matrix of the radar sensor obtained is:

$$R_{k-R} = \begin{bmatrix} 0.088 & 0 & 0 & 0 & 0 \\ 0 & 0.00088 & 0 & 0 & 0 \\ 0 & 0 & 0.088 & 0 & 0 \\ 0 & 0 & 0 & 0.0088 & 0 \\ 0 & 0 & 0 & 0 & 0.08 \end{bmatrix} \tag{33}$$

For this matrix, the covariance values are directly related to the resolution and variability around the measurements of the LiDAR and Radar sensors previously characterized for a generic situation. The purpose of this chapter is to investigate the response of the EKF and UKF filter in case of significant distortion increase of the x and y measurements for the LiDAR case and magnitude ρ, angle θ, and target speed in the Radar case.

About the implementation, the EKF includes the nonlinear functions sine and cosine, and the CORDIC (Coordinate Rotation Digital Computer) solves mathematical operations from three parameters, which allows to efficiently obtain calculations of sinusoidal functions, exponential, logarithms, roots, inverse trigonometric functions, and hyperbolic functions using additions, iterations and rotations of bits, this rotation process always converges in a fixed number of iterations [8].

2.1 Response of EKF

The performance of the EKF was examined by adding a contaminated signal and comparing the actual value with the filter response. For the CTRV model, the tracking of mobile objects by merging Lidar and Radar data achieves a good approach to the real measurement, counteracting the Gaussian noise associated with the uncertainty in the measurement of a sensor. The following **Figure 2** shows the filter response.

To verify the filter's ability to track targets, real measurements have been taken from lidar and radar sensors, for a predetermined trajectory shown by the blue line in **Figure 2**. The x and y positions at a height z are our variables of interest, the filter response represented by the red line, for the first 450 samples evidences the reduction in the scattering of the sensor data around the real trajectory.

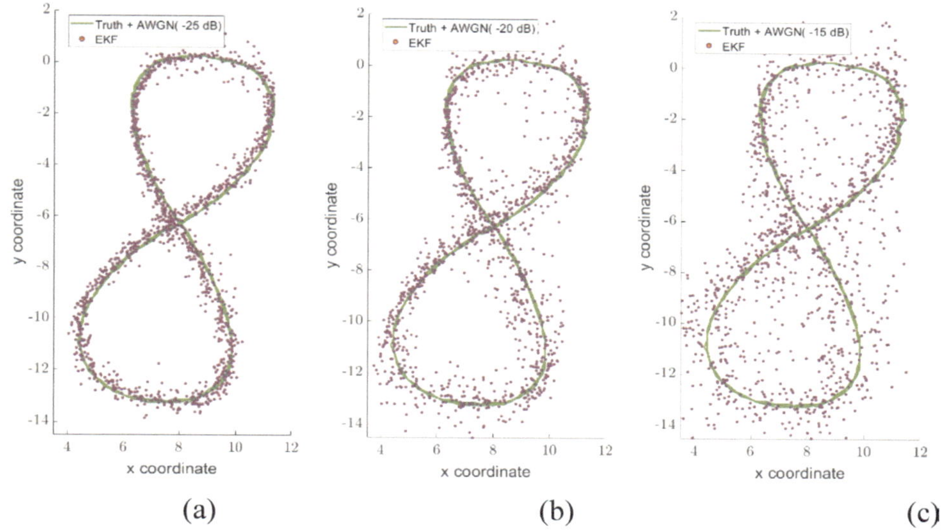

Figure 4.
EKF response when contaminating the dataset with AWGN (MATLAB) for a. −25 dB, b. −20 dB and c. −15 dB.

To evaluate the robustness of the filter, a dataset has been built using the same scene by capturing cloud points with simultaneous Lidar and Radar data. For the Lidar signal, the data provides the x and y position in rectangular coordinates, and for the Radar case, the data has the magnitude and angle of the detected object and the magnitude of the detected angular velocity of the tracked object. Besides, this data set is degraded by adding white Gaussian noise (AWGN), and by considering that not all sensors provide good accuracy in their measurements, an evaluation of the response of the EKF when the data obtained from the sensors undergo progressive contamination is performed with results as shown **Figure 4**.

Graphically the real trajectory for **Figure 4** is shown in green and the EKF estimates are in black. **Figure 4**. *a* shows the filter estimation for the measurements degraded by noise and with zero gain; for **Figure 4**. *b* a gain of 15 dB is included, and for **Figure 4**. *c* the gain increases up to 30 dB. The parameter to quantify the ability of the filter to discriminate the noise and approach the real values is the root mean square error (RMSE), the comparison of this parameter for the response of the EKF and the UKF will be examined later. A low capability of the EKF to reject high levels of noise associated with Lidar and Radar sensor measurements is evident.

3. UKF model

The UKF filter uses the UT (Unscented transformation) to improve the estimates of the first moments of a first random variable [9–11] by propagating a second Gaussian random variable through a nonlinear transformation [12]. This allows the UKF not only to have a higher convergence speed of the filter but also to always guarantee this convergence. In order to take the CTRV model to three dimensions, the following assumptions were taken into account:

Having x_k, y_k state vectors, u_k work variables, v_k noise associated with the process, and n_k noise associated with the measurements for an instant in time k [11, 13], the UKF filter starts from the following model:

$$\mathbf{x}_k = F(\mathbf{x}_{k-1}, \mathbf{u}_{k-1}, \mathbf{v}_{k-1}) \tag{34}$$

For the proposed model using data sensor fusion, the drone-mobile object scenario and its dynamic model can be described as:

$$f = P_x + V_x \cdot \Delta T \tag{35}$$

$$f = P_y + V_y \cdot \Delta T \tag{36}$$

Where the fundamental matrix corresponds to:

$$F = \begin{bmatrix} 1 & 0 & \Delta T & 0 \\ 0 & 1 & 0 & \Delta T \\ 0 & 0 & 1 & 0 \\ 0 & 0 & 0 & 1 \end{bmatrix} \tag{37}$$

The vector of measurements in three dimensions can be expressed as [14]:

$$\mathbf{y}_k = H(\mathbf{x}_k, \mathbf{n}_k) \tag{38}$$

$$H = \begin{bmatrix} \sqrt{{P_x}^2 + P_y^2 + P_z^2} \\ \tan^{-1} \dfrac{P_y}{P_x} \\ \dfrac{P_x \cdot V_x + P_y \cdot Vy}{\sqrt{{P_x}^2 + P_y^2 + P_z^2}} \end{bmatrix} \tag{39}$$

Initialization: One of the characteristics of the UKF is the concatenation in the vector of states and the covariance matrix, which will be denominated as $x_a(k\text{-}1)$ and $P_a(k\text{-}1)$, therefore they are initialized by fulfilling that:

$$\hat{\mathbf{x}}_{k-1}^{\mathrm{a}} = [\mathbf{x}_k \;\; \mathbf{v}_k \;\; \mathbf{n}_k]^T \tag{40}$$

$$\mathbf{P}_{k-1}{}^{\mathrm{a}} = \begin{bmatrix} \mathbf{P}_{k-1|k-1} & 0 & 0 \\ 0 & \mathbf{Q} & 0 \\ 0 & 0 & \mathbf{R} \end{bmatrix} \tag{41}$$

With Q and Rr being the following matrix:

$$\mathbf{Q} = \begin{bmatrix} \dfrac{\Delta T^2}{4} & 0 & \dfrac{\Delta T^3}{2} & 0 \\ 0 & \dfrac{\Delta T^2}{4} & 0 & \dfrac{\Delta T^3}{2} \\ \dfrac{\Delta T^3}{2} & 0 & \Delta T^2 & 0 \\ 0 & \dfrac{\Delta T^3}{2} & 0 & \Delta T^2 \end{bmatrix} \tag{42}$$

$$R_r = \begin{bmatrix} 0.09 & 0 & 0 \\ 0 & 0.005 & 0 \\ 0 & 0 & 0.09 \end{bmatrix} \tag{43}$$

Sigma points are defined as:

$$X^{\mathrm{a}}_{k-1} = \left[\hat{\mathbf{x}}^{\mathrm{a}}{}_{k-1} \quad \hat{\mathbf{x}}^{\mathrm{a}}{}_{k-1} + \sqrt{(L+\lambda)\cdot \mathbf{P}_{k-1}{}^{\mathrm{a}}} \quad \hat{\mathbf{x}}^{\mathrm{a}}{}_{k-1} - \sqrt{(L+\lambda)\cdot \mathbf{P}_{k-1}{}^{\mathrm{a}}} \right] \tag{44}$$

The update of the sigma points through the F function is given by:

$$x^{\mathrm{a}}_{k|k-1} = F\left[x^{\mathrm{a}}_{k-1}, \mathbf{u}(k)\right] \tag{45}$$

Calculating $\hat{\mathrm{x}}_{\mathrm{k-1}}$ and P_{k-1} is possible by means of the following expressions:

$$\hat{\mathbf{x}}_k^- = \sum_{i=0}^{2L} W_i^m X_{i,k|k-1} \tag{46}$$

$$\mathbf{P}_k^- \approx \sum_{i=0}^{2L} W_i^{(c)} \left[X_{i,k|k-1} - \hat{\mathbf{x}}_k^-\right]\left[X_{i,k|k-1} - \hat{\mathbf{x}}_k^-\right]^T \tag{47}$$

In the same way, the vector of innovations of the process is also determined:

$$\mathbf{I}_{k|k-1} = H\left[x_{k|k-1}, \mathbf{u}(k)\right] \tag{48}$$

$$\hat{\mathbf{y}}_k^- = \sum_{i=0}^{2L} W_i^m \mathbf{I}_{i,k|k-1} \tag{49}$$

Updating with measurements: the covariance matrix of the measurements is calculated as:

$$\mathbf{P}_{y_k,y_k} \approx \sum_{i=0}^{2L} W_i^{(c)} \left[\mathbf{I}_{i,k|k-1} - \hat{\mathbf{y}}_k^-\right]\left[\mathbf{I}_{i,k|k-1} - \hat{\mathbf{y}}_k^-\right]^T \tag{50}$$

The cross-correlation between the estimated states and the sequence of measurements is:

$$\mathbf{P}_{x_k,y_k} \approx \sum_{i=0}^{2L} W_i^{(c)} \left[X_{i,k|k-1} - \hat{\mathbf{x}}_k^-\right]\left[\mathbf{I}_{i,k|k-1} - \hat{\mathbf{y}}_k^-\right]^T \tag{51}$$

Then, it is necessary to evaluate the gain of the Kalman filter and update the $\hat{\mathrm{x}}_{\mathrm{k}}$ and P_{k-1}

$$\mathbf{K} = \mathbf{P}_{x_k,y_k} \mathbf{P}^{-1}_{y_k,y_k} \tag{52}$$

$$\hat{\mathbf{x}}_k = \hat{\mathbf{x}}_k + \mathbf{K}\left(\mathbf{y}_k - \hat{\mathbf{y}}_k^-\right) \tag{53}$$

$$\mathbf{P}_k = \mathbf{P}_k^- + \mathbf{K}\cdot \mathbf{P}_{y_k,y_k} \cdot \mathbf{K}^T \tag{54}$$

3.1 Response of UKF

Similar to the way of evaluating the robustness of the EKF, the UKF its response has been subjected to a dataset also contaminated with AWGN in a stepwise manner, the response of the UKF is presented in the **Figure 5**.

Despite not having a modeling of UAV dynamics and mobile targets, the UKF achieves a very good rejection of noise associated with the data obtained from the Lidar and Radar sensors, managing to estimate to a large extent the real trajectory.

Figure 6 shows the strong ability of the UKF to reject high levels of noise associated with Lidar and Radar sensor measurements is evident. In real practice, the uncertainty associated with the measurements can be increased by different factors: sensor failures, very aggressive measurement environments, etc. For this reason, the UKF presents a more robust response to severe degradations in the data obtained from the sensors (**Figure** 7).

When performing a sweep in the gain of the noise added to the measurements of the Lidar and Radar sensors, and comparing the response of the EKF and UKF taking the RMSE as a reference parameter, the following response was observed.

Figure 8 shows a better capability of the UKF filter with respect to the EKF, for rejection of high noise levels in the sensor measurements, even when the noise level keeps increasing, the UKF response maintains low RMSE levels, i.e. it achieves a good estimation with respect to the real target position values. The following **Table 1** shows the RMSE and MSE values for UKF and EKF with increasing AWGN. Note that the UKF response improves as the AWGN is increased.

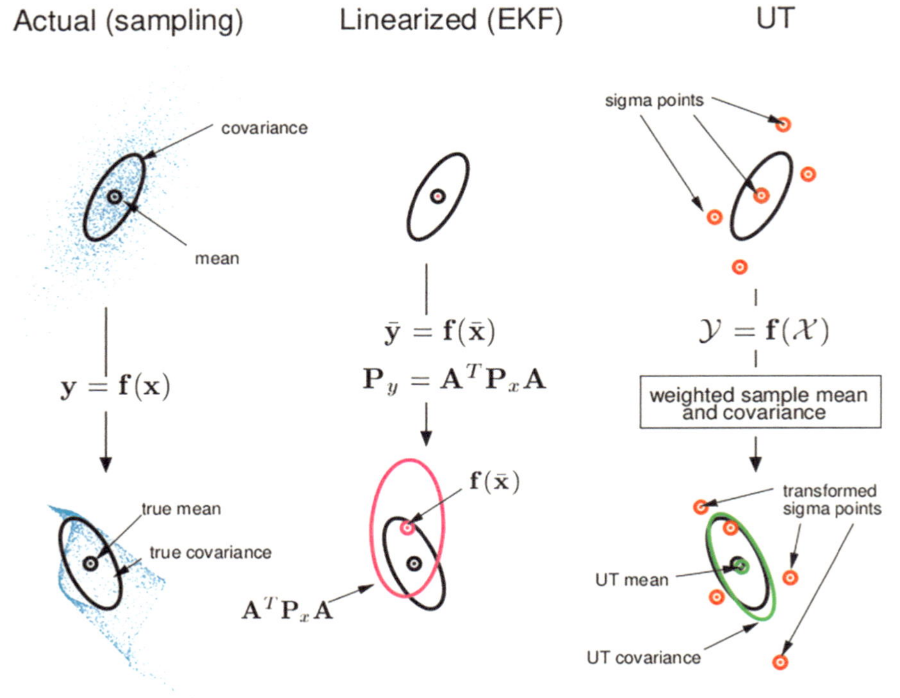

Figure 5.
Example of the UT for mean and covariance propagation. Taken from [12].

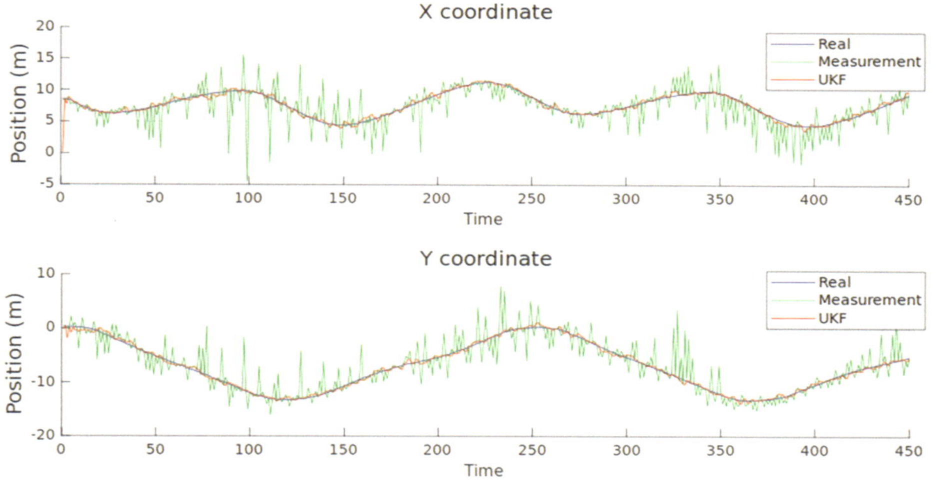

Figure 6.
EKF performance for x and y position estimation.

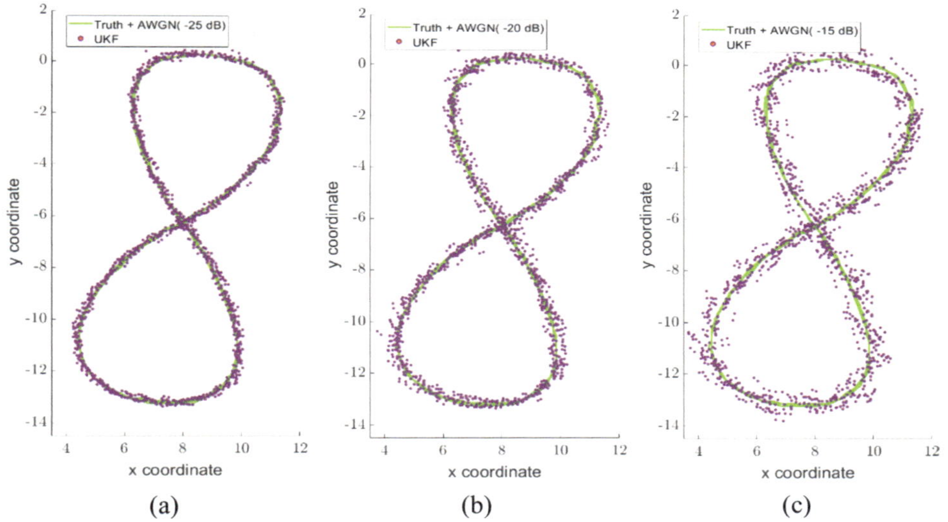

Figure 7.
UKF response when contaminating the dataset with AWGN (MATLAB) for a. −25 dB, b. −20 dB and c. −15 dB.

Percentageally for −25 dB the UKF filter reduces noise by 14.8% more than the EKF filter, as the noise intensity is greater, this difference also increases with a value of 27.72% for −20 dB, and 38.23% at −15 dB.

4. Conclusions

For a correct filtering with EKF, good modeling of the system dynamics is required to achieve an accurate a priori prediction in the first stage of the filtering since the convergence of this filter by means of the Kalman gain is slower than the convergence of the UKF.

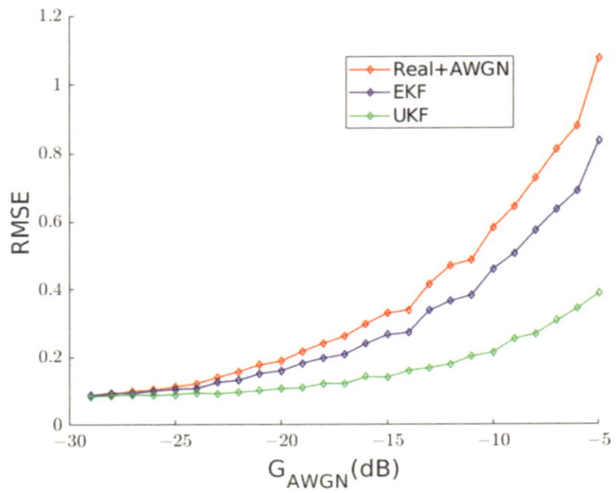

Figure 8.
RMSE for EKF and UKF with increasing noise gain added to the dataset.

RMSE			
G_AWGN	COORDINATE	EKF	UKF
−25dB	Px	0.1053	0.089
	Py	0.125	0.1
−20dB	Px	0.159	0.1073
	Py	0.1649	0.1166
−15dB	Px	0.266	0.14
	Py	0.2582	0.147
MSE			
G_AWGN	**COORDINATE**	**EKF**	**UKF**
−25dB	Px	0.011	0.008
	Py	0.016	0.01
−20dB	Px	0.025	0.012
	Py	0.027	0.014
−15dB	Px	0.071	0.02
	Py	0.067	0.022

Table 1.
RMSE and MSE for X coordinate and Y coordinate.

The UKF has a higher computational complexity compared to the EKF, when the CTRV model is combined with this last filter, a reduction of the computational complexity is obtained when it is implemented together with CORDIC.

When there is a fusion of Lidar and Radar sensor data in a drone with omnidirectional motion capability, the UKF filter has a better rejection of the noise associated with the measurements than the UKF filter.

Even with limited modeling of the system dynamics, the UKF manages to obtain an excellent response in the rejection of the noise associated with the fusion of the sensors.

By employing a progressive and constant increase of the distortion in the data obtained from the Lidar and Radar sensors, it was observed that the UKF has a more robust response to the increase of the uncertainty of the process to be filtered. For the actual implementation of a sensor data fusion system, this indicates a better adaptability of the UKF filter to different measurement scenarios as well as to differences in the quality of the sensors used.

Acknowledgements

The UPTC is part of the International Alliance with the Istituto di Tecnologie della Comunicazione dell'Informazione e della Percezione (TeCIP) from Italia and the Ljubljana University of Slovenia. We acknowledge NATO for the funds that make possible the research under grant G-5888 Science for Peace - Clarifier Research Project, and also, thanks to the UPTC for the administration of the funds under SGI-3139.

Conflict of interest

The authors declare no conflict of interest.

Author details

Oscar M. Sogamoso*, Eduardo A. Fernández and Marco J. Suarez
University Pedagogical and Technological of Colombia, Sogamoso, Colombia

*Address all correspondence to: oscarjavier.montanez@uptc.edu.co

References

[1] Cesviweb_user. Los sistemas ADAS sí reducen la accidentalidad [Internet]. 2017. Available from: https://www.cesvicolombia.com/los-sistemas-adas-si-reducen-la-accidentalidad/

[2] Lee H, Chae H, Yi K. A geometric model based 2D LiDAR/radar sensor fusion for tracking surrounding vehicles. IFAC-PapersOnLine. 2019;**52**(8): 277-282. DOI: 10.1016/j.ifacol.2019.08.060

[3] Kim B, Yi K, Yoo HJ, Chong HJ, Ko B. An IMM/EKF approach for enhanced multitarget state estimation for application to integrated risk management system. IEEE Transactions on Vehicular Technology. 2015;**64**(3): 876-889. DOI: 10.1109/TVT.2014.2329497

[4] Farag W. Kalman-filter-based sensor fusion applied to road-objects detection and tracking for autonomous vehicles. Proceedings of the Institution of Mechanical Engineers, Part I: Journal of Systems and Control Engineering. 2021; **235**(7):1125-1138. DOI: 10.1177/0959651820975523

[5] Wang Y, Liu D, Matson E. Accurate Perception for Autonomous Driving: Application of Kalman Filter for Sensor Fusion. In: 2020 IEEE Sensors Applications Symposium (SAS); 09-11 March 2020; Kuala Lumpur. Malaysia: IEEE; 2020. p. 1-6, DOI: 10.1109/SAS48726.2020.9220083

[6] Gandolfo DC, Salinas LR, Brandao A, Toibero JM. Stable path-following control for a quadrotor helicopter considering energy consumption. IEEE Transactions on Control Systems Technology. 2017;**25**(4):1423-1430. DOI: 10.1109/TCST.2016.2601288

[7] Montañez OJ, Suarez MJ, Fernandez EA. Application of data sensor fusion using extended Kalman filter algorithm for identification and tracking of moving targets from LiDAR–radar data. Remote Sensing. 2023;**15**: 3396. DOI: 10.3390/rs15133396

[8] Volder JE. The CORDIC trigonometric computing technique. IRE Transactions on Electronic Computers. 1959;**EC-8**(3):330-334. DOI: 10.1109/TEC.1959.5222693

[9] Tian K, Radovnikovich M, Cheok KC. Comparing EKF, UKF, and PF performance for autonomous vehicle multi-sensor fusion and tracking in highway scenario. In: IEEE SysCon 2022: The 16th Annual IEEE International Systems Conference Proceedings; 25-28 April 2022; Montreal. Canada: IEEE; 2022. p. 1-6. DOI: 10.1109/SysCon53536.2022.9773872

[10] Bucci A, Franchi M, Ridolfi A, Secciani N, Allotta B. Evaluation of UKF-based fusion strategies for autonomous underwater vehicles multisensor navigation. IEEE Journal of Oceanic Engineering. 2023;**48**(1):1-26. DOI: 10.1109/JOE.2022.3168934

[11] Choi M, Seo M, Kim HS, Seo T. UKF-based sensor fusion method for position estimation of a 2-DOF rope driven robot. IEEE Access. 2021;**9**:12301-12308. DOI: 10.1109/ACCESS.2021.3051404

[12] Eric W, der Merwe V. The unscented kalman filter. In: Simon H, editors. Kalman Filtering and Neural Networks. 1st ed. Chichester: Wiley; 2001. pp. 221-280. DOI: 10.1002/0471221546

[13] Jeon D, Choi H, Kim J. UKF data fusion of odometry and magnetic sensor

for a precise indoor localization system of an autonomous vehicle. In: 2016 13th International Conference on Ubiquitous Robots and Ambient Intelligence; 19-22 August 2016; Xi'an. China: IEEE; 2016. pp. 47-52. DOI: 10.1109/URAI.2016.7734018

[14] Yuan Z, Falin W, Yushuang L, Zhidong Z. IMM-UKF Based Airborne Radar and ESM Data Fusion for Target Tracking. In: 2019 14th IEEE International Conference on Electronic Measurement & Instruments (ICEMI); 01-03 November 2019; Changsha. China: IEEE; 2019. pp. 586-592. DOI: 10.1109/ICEMI46757.2019.9101826

Section 2

Optimizations and Enhancements for Data Acquisition, Target Tracking, and Pairs Trading

Chapter 4

Perspective Chapter: Optimizing µ-PMU Placement for Estimating Asymmetrical Distribution Network States – Introducing a Novel Stochastic Two-Stage Approach

Arya Abdolahi, Selma Cheshmeh Khavar,
Morteza Nazari-Heris and Navid Taghizadegan Kalantari

Abstract

Distribution system state estimation plays a crucial role in supplying essential data for system monitoring and control. However, the presence of uncertain parameters such as the variable output of distributed generation (DG), random meter errors, and inaccurate network parameters poses a significant challenge to achieving situational awareness (SA) in distribution systems. To address these challenges, this study introduces an innovative two-stage stochastic programming model. In the first stage, the model focuses on the optimal placement of µ-PMUs with the objective of minimizing installation costs, maximizing measurement redundancy, and enhancing system observability. This is particularly important in the context of partially zero injection nodes (PZIN). In the second stage, the model performs state estimation for three-phase asymmetric DG-integrated distribution systems, aiming to improve SA. The application of the proposed model resulted in the identification of optimal µ-PMU locations in the presence of PZINs and various contingencies. The distribution system state estimation achieved high accuracy and a low error percentage. The feasibility and effectiveness of this methodology were validated using the modified IEEE 85-bus distribution system. In addition, we incorporated extended Kalman filter (EKF) state estimation to compare with the weighted least square method.

Keywords: distribution system state estimation, µ-Phasor measurement unit, distributed generation, observability, asymmetrical structure

 IntechOpen

1. Introduction

1.1 Concept and incitement

Monitoring in real-time is crucial for implementing control and protection functions in an electric distribution network. Efficiently leveraging current measurements is essential for producing the most precise estimation of the system state that corresponds to the available data. Distribution System State Estimation (DSSE) is acknowledged as a critical component of the distribution management system. Despite significant progress in TSSE over time, it is essential to recognize the fundamental differences between transmission and distribution networks. Consequently, the functional needs of TSSE cannot be directly translated to DSSE [1]. Distinctive features of distribution systems encompass load imbalances, elevated R/X ratios, and reduced observability stemming from inadequate real-time monitoring. On the other hand, transmission systems possess a surplus of measurements, ensuring ongoing system observability through redundancy [2]. Observability in distribution systems faces constraints due to the restricted availability of real-time measurements. The incorporation of μ-PMUs into distribution systems improves the accuracy of DSSE by furnishing precise details on voltage magnitude and phase angle. In today's distribution networks, the performance requirements for DSSE are becoming more rigorous, primarily driven by challenges stemming from the integration of DG and the adoption of sophisticated technology. The main impediments and difficulties linked to DSSE are concisely delineated in [3].

In transmission networks, the inherent assumption is the operation's symmetry, necessitating all calculations to be performed on single-phase positive sequences [4]. Distribution systems possess unique attributes, such as a radial structure, unbalanced loads, and a deficiency in real-time measurements, distinguishing them from transmission systems. In situations like these, the application of state estimation methods originally tailored for transmission systems is impractical for distribution systems, necessitating the adoption of three-phase state estimation. The scarcity of real-time measurements is a distinctive characteristic of DSSE. To ensure system observability, the incorporation of pseudo measures becomes necessary, representing the active and reactive powers of both loads and distributed generation (DG) [5].

1.2 Literature review

The authors in Refs. [6–11] propose several approaches to modeling the mathematical problem, application of μ-PMUs, and locating the metering instruments utilized in Distribution System State Estimation (DSSE). In [12], a pioneering hybrid approach was introduced for state estimation in power systems. This method combines weighted least squares with the integration of phasor measurement units and supervisory control and data acquisition systems. In Ref. [13], a novel algorithm is presented for intricate linear state estimation. This algorithm considers the noise parameter and incorporates data from distribution phasor measurement units to improve the SA of distribution systems. Utilizing high-quality data from these units, this method facilitates the observability of unbalanced and dynamic distribution systems. Regularized state estimation technique was developed in Ref. [14] for robust monitoring of the distribution network. The primary objective was to achieve precise tracking of the system state on a faster time scale, thereby enhancing reliability according to the needs of the new operating environment. Ref. [15] introduces a method designed to enhance the reliability of the state estimation process and equipment performance within power system substations. This involves strategically situating channels and sources of phasor measurement units. The method consists of two stages. In the first stage, the application of a genetic algorithm optimizes the placement of phasor measurement units. Following this, the second stage employs an innovative approach to optimize measurements and allocate channels effectively.

Authors in paper [16] introduce an innovative approach for evaluating the condition of unbalanced three-phase distribution systems. This methodology utilizes branch currents as state variables, calculated in rectangular coordinates. Employing Phasor Measurement Units (PMUs) located at monitored busses, the approach provides branch currents and nodal voltages. Inequality constraints in DSSE involve non-monitored load busses, constrained by daily load variations from the preceding estimation period. The approach proposed by the authors in [17] innovatively combines the weighted least squares (WLS) and Levenberg-Marquardt methods. This fusion relies on incorporating data sourced from smart meters, significantly contributing to the implementation of real-time voltage control strategies. A novel state estimation framework has been introduced in [18], incorporating equality constraints related to voltage-dependent loads and zero injection. This framework aims to improve both the accuracy of state estimation and the effectiveness of bad data detection. Reference [19] introduces a unique method for static harmonic state estimation in distribution networks. This approach utilizes optimization techniques and assumes the strategic positioning of a small number of phasor measurement devices along the feeders. In [20], the author advocates a data-driven strategy to correct measurement errors inherent in the traditional weighted least squares methodology.

Authors in paper [21] present an innovative decentralized state estimation approach employing a multi-agent system to address the intricacies of distribution systems. The key idea involves dividing the system into smaller subsystems for more efficient and rapid estimation. Subsequently, a metaheuristic algorithm, specifically the artificial bee colony algorithm, is applied to solve the state estimation problem in a multi-agent context. Reference [22] introduces an original framework for the multi-objective optimization of Phasor Measurement Unit (PMU) placement. The goal is to minimize the total PMU installation cost, optimize the selection of current channels, and diminish errors in state estimation. To address this mixed-integer nonlinear programming problem, the non-dominated sorting genetic algorithm II is employed as the solution approach. Reference [23] introduces a comprehensive method for estimating and monitoring the overall state of low-voltage distribution networks, specifically those with significant integration of photovoltaic systems. The effectiveness of the proposed technique is showcased in scenarios with high imbalance. Validation of the efficiency of the suggested low-voltage linear state estimation method is carried out through testing on a representative imbalanced residential network. A comprehensive modeling strategy, as suggested in [24], focuses on integrating the effects of Zero-Injection Nodes (ZIN) to optimize the positioning of Phasor Measurement Units (PMUs) and guarantee full observability within power transmission systems. Achieving globally optimal solutions involves the utilization of an intelligent BTS algorithm in the presentation of optimal PMU placement strategies. In Ref. [25], an optimal placement model for Phasor Measurement Units (PMUs) in transmission networks is presented. This model, utilizing multi-objective mixed-integer linear programming, considers variables including the installation cost of

PMUs, observability constraints, and the capability to identify gross errors. A framework for investigating gross errors is introduced in [26], which employs μ-PMUs to separate measurement errors. This approach involves analyzing gross errors in the measurements acquired from both smart meters and SCADA devices as a post-processing step within the non-linear DSSE framework.

The measurements of both active and reactive power injection demonstrate non-linear features, without a readily available direct solution. The determination of the measurement function requires an iterative approach, as highlighted in Ref. [27]. Asymmetric distribution systems and the inherent non-linear nature of measurement data prompt the formulation of models in this research field as non-linear programming problems. Diverse solvers are employed to tackle and resolve challenges posed by these problems. A variety of potent metaheuristic algorithms are employed to attain globally optimal solutions, including a combination of particle swarm optimization and chaotic gravitational search algorithms [5], genetic algorithm [15], artificial bee colony algorithm [21], non-dominated sorting genetic algorithm-II [22], and a hybrid multi-objective particle swarm optimization krill herd algorithm [28]. Established solvers such as Gurobi and CPLEX [29] are applied to address challenges in DSSE and obtain accurate solutions for the system state. This study introduces a DSSE model aimed at improving observability and reducing computational complexity and calculation time in DG-based distribution networks. Employing the Taylor series approach, this model approximates and linearizes the non-linear functions related to active and reactive power, leading to a more thorough state estimation compared to existing DSSE techniques. The final step involves computing the estimated state vector using the weighted least square method.

1.3 Novelties and contribution

Despite notable advancements in DSSE, there remain several challenges and deficiencies that require careful consideration. In brief, the inadequacies identified in previous references can be summarized as follows:

a. In distribution networks, there is a multitude of distributed smart meters; however, only a limited subset of these devices can simultaneously sample voltage and power values. Additionally, the data collection process for these devices, reliant on wireless signals for data transfer to an access point, is susceptible to data loss. Given these practical challenges, the quantity of measurements in distribution networks falls significantly short of ensuring system observability.

b. Distribution networks with asymmetrical structures exhibit distinct characteristics, including the presence of single-phase and two-phase loads at specific busses. Current research employs Fully Zero Injection Nodes to minimize the need for μ-PMUs. In the context of FZIN, a bus in an asymmetrical structure catering exclusively to a single-phase load is designated as a non-zero injection bus. It's important to note that two phases of this bus exhibit zero injection characteristics, potentially enhancing network observability with a reduced number of μ-PMUs. However, relying solely on FZIN may not fully demonstrate the effectiveness of zero injection properties.

To address the shortcomings identified in prior literature, this chapter presents an innovative two-stage programming model. In the first phase, the emphasis is on resolving the optimal μ-PMU placement problem. The objective is to minimize the installation cost of μ-PMUs while ensuring extensive observability in the presence of PZIN under various contingencies. The second phase involves conducting a three-phase asymmetrical DSSE to enhance SA. The principal novel contributions of this chapter are delineated as follows:

1. To overcome the first limitation (a), this paper presents a DSSE model that employs the Taylor series approach for approximations. The objective is to linearize the non-linear functions linked to active and reactive power, ensuring an exhaustive state estimation of the network that exceeds the capabilities of existing DSSE methods.

2. To address the second limitation (b), the proposed optimal μ-PMU placement problem incorporates the concept of PZINs. This concept pertains to busses with solely one or two zero injection phases, and it is employed to improve the overall observability of three-phase asymmetrical distribution networks.

Besides the key advancements mentioned previously, this paper introduces several contributions outlined as follows:

i. Utilizing both single-phase and three-phase μPMUs in radial distribution networks with asymmetrical structures to minimize the required number of installed μ-PMUs. To implement this concept, a conventional model for the optimal placement of PMUs in symmetrical transmission networks is adapted, and a distinct model is derived. In this unique model, the objective function and constraints align with the asymmetrical characteristics of radial distribution networks.

ii. Considering various contingency situations in distribution networks, including the failure of a single μPMU and an individual line outage, within the model designed for the asymmetrical and radial structure of distribution networks.

iii. The proposed approach is formulated as a MILP problem, and the CPLEX solver is utilized to minimize and achieve the best global solution.

iv. Incorporating the optimal placement of μ-PMUs and DSSE issues into a two-stage programming problem, where the decision variables determining the optimal μ-PMU placement (first stage) serve as input data for the DSSE (second stage).

1.4 Chapter organization

A comparative analysis of Kalman filters and WLS for state estimation in distribution systems is presented in Section 2. In Section 3, we outline the problem formulation, consisting of two stages. The first stage focuses on the placement of μ-PMUs, while the second stage is dedicated to addressing state estimation. Section 4 provides a summary of the solver utilized in the context of two-stage stochastic programming. The outcomes of simulations and associated discussions are detailed in Section 5. Ultimately, Section 6 encapsulates the concluding remarks.

2. Comparative analysis of EKF and WLS for state estimation

Among the numerous state estimation techniques available, Kalman filters and WLS are widely used in distribution systems for their robustness and accuracy. This discussion aims to compare these two techniques, examining their principles, applications, and evaluating their performance in different scenarios.

2.1 Kalman filter

Kalman filters are recursive algorithms designed for state estimation in dynamic systems. They operate on a prediction-correction cycle, combining a system model with measurements to update the state estimate. In distribution systems, Kalman filters are commonly employed for real-time tracking and estimation of variables such as voltage, current, and power flow. The applications of Kalman filters in distribution systems are listed as follow:

1. *Power system state estimation:* Kalman filters excel in tracking the state of dynamic power systems, handling uncertainties and noise in measurements effectively.

2. *Sensor fusion:* Kalman filters are employed for integrating data from various sensors in distribution systems, enhancing the accuracy of state estimation.

3. *Fault detection:* Kalman filters can be used to identify faults in distribution systems by analyzing deviations between predicted and measured values.

2.2 WLS method

WLS is a statistical method used for estimating unknown parameters by minimizing the sum of squared weighted residuals. In the context of distribution systems, WLS is often applied to linear regression problems, making it suitable for static state estimation. The application of WLS in distribution systems are listed as follow:

1. *Load flow analysis:* WLS is utilized for estimating the state variables in load flow studies, providing a reliable representation of the system's current operating conditions.

2. *Bad data detection:* WLS helps identify and mitigate the impact of erroneous measurements in distribution systems, improving the overall accuracy of state estimation.

3. *Parameter estimation:* Weighted Least Squares is used for estimating parameters in distribution system models, aiding in system planning and optimization.

2.3 Comparative analysis

1. *Computational complexity:* Kalman filters typically involve more complex computations due to their recursive nature and reliance on dynamic models. WLS, being a static technique, generally has lower computational requirements.

2. *Robustness to Non-linearity:* Kalman filters are well-suited for nonlinear systems, while WLS is more effective in linear scenarios. The choice between the two depends on the nature of the distribution system and the extent of non-linearity.

3. *Handling of measurement noise:* Kalman filters are designed to handle measurement noise more effectively than WLS, making them preferable in applications where accurate tracking of dynamic variables is crucial.

3. Problem formulation

3.1 Extended Kalman filter formulation

The EKF consists of two primary stages: the a priori phase, which involves prediction or processing based on the previous state estimate computed in the prior iteration, indicated by superscript "−", and the a posteriori phase, involving measurement update or estimation. These estimates are computed before any system measurements are taken.

The EKF operates under specific assumptions:

- Gaussian and uncorrelated noise characterize both the measurement and process.

- The system is observable, as defined mathematically below:

$$x_k = g(u_k, x_{k-1}) + \varepsilon_k \tag{1}$$

$$z_k = h(x_k) + \delta_k \tag{2}$$

Here, $\in_k$ represents the process noise and δ_k denotes the measurement noise, both having a zero mean. The functions g and h necessitate linearization through the computation of their Jacobian matrices.

In every time step (k), the EKF algorithm defines the estimation x_k through its mean μ_k and covariance $\sum_k$. EKF follows a recursive process with two sequential steps, continuing until an accurate estimation is reached. The determination of both the mean and covariance occurs through separate steps:

1. Prediction

$$\bar{\mu}_k = g(u_k, \mu_{k-1}) \tag{3}$$

$$\bar{\sum_k} = G(u_k, \mu_{k-1}) \sum_{k-1} G^T(u_k, \mu_{k-1}) + Q \tag{4}$$

2. Correction

$$K_k = \bar{\sum_k} H^T(\bar{\mu}_k)\left[H(\bar{\mu}_k)\bar{\sum_k} H^T(\bar{\mu}_k) + R\right]^{-1} \tag{5}$$

$$\mu_k = \bar{\mu}_k + K_k[z_k - h(\bar{\mu}_k)] \tag{6}$$

$$\sum_k = [I - K_k H(\bar{\mu}_k)]\bar{\sum_k} \tag{7}$$

where $g(u_k, \mu_{k-1})$ is linearized around the mean μ_{k-1}, $G(u_k, \mu_{k-1}) = \frac{(u_k, \mu_{k-1})}{\delta x_{k-1}}\Big|_{x_{k-1}=\mu_{k\text{-}1}}$ and Q is the covariance of the noise matrix. The matrices H, h and R are the same as in the WLS method. The measurement z_k is only incorporated in the algorithm during the correction step. K_k represents the Kalman gain, determining how much the new state estimate is influenced by measurements. Thus, each iteration integrates a fresh set of measurements for correction. However, it's reasonable to assume that the average remains steady, given the power system's stability across consecutive time steps. Therefore, (3) becomes:

$$\bar{\mu}_k = \mu_{k-1} \tag{8}$$

Thus, it follows that g is independent from u_k and $G(u_k, \mu_{k-1}) = \frac{(u_k, \mu_{k-1})}{\delta x_{k-1}}\Big|_{x_{k-1}=\mu_{k\text{-}1}}$ becomes an identity matrix. The updated EKF formulation is:

1. Prediction

$$\bar{x}_k = x_{k-1} \tag{9}$$

$$\bar{\sum_k} = \sum_{k-1} + Q \tag{10}$$

2. Correction

$$K_k = \bar{\sum_k} H^T(\bar{x}_k)\left[H(\bar{x}_k)\bar{\sum_k} H^T(\bar{x}_k) + R\right]^{-1} \tag{11}$$

$$x_k = \bar{x}_k + K_k[z_k - h(\bar{x}_k)] \tag{12}$$

$$\sum_k = [I - K_k H(\bar{x}_k)]\bar{\sum_k} \tag{13}$$

3.2 μ-PMU placement problem

For the purpose of SA and monitoring, PMUs can supply precise, high-resolution, and directly synchronized phasor measurements [30, 31]. The development of μ-PMUs, utilizing the same synchrophasor technology, is aimed at improving the intelligent operation of power distribution systems, thereby enabling their observability. When compared to siting PMUs in high-voltage transmission networks, μ-PMUs prove to be significantly more cost-effective for widespread deployment in distribution systems. This may potentially address the challenge of limited observability. The subsequent formulation is presented to ascertain the optimal allocation of μ-PMUs in asymmetrical systems.

3.2.1 Objective function

In the first stage, the objective function pertaining to the allocation of μ-PMUs is defined by Eq. (14). This specified objective consists of two elements, where WX_j and VP_j are associated with the three-phase and single-phase aspects, respectively. The objective is to identify the most favorable network locations for the installation of μ-PMUs, with the overarching aim of minimizing the total investment cost while maximizing the observability index of the distribution system. In subsequent sections, various contingencies, including PZIN, are integrated into this objective function.

$$\text{Min} \sum_{j \in \Omega_B} \sum_{n \in \{a,b,c\}} WX_j + VP_j^n \tag{14}$$

$$3X_j + 2 \times \left[P_j^a + P_j^b + P_j^c\right] \leq 4 \;\; j = 1, \ldots, \text{N} \tag{15}$$

Eq. (15) guarantees that each bus is limited to having either a three-phase μ-PMU or, at most, two single-phase μ-PMUs installed.

3.2.2 Observability constraints under the normal operating conditions

The constraint on topological observability, influenced by both the location of μ-PMUs and zero injection properties, is formulated as Eqs. (16)–(18) [32, 33].

$$f_i = \sum_{j \in \Omega_B} \sum_{n \in \{a,b,c\}} a_{ij}^n XX_j^n + \sum_{j \in \Omega_B} \sum_{n \in \{a,b,c\}} a_{ij}^n Y_{ij}^n Z_j^n \geq 3 - \mathit{off}_i^n \tag{16}$$

$$XX_j^n = X_j + P_j^n \;\; i,j = 1, \ldots, \text{N}, \;\; \text{n} \in \{a,b,c\} \tag{17}$$

$$\sum_{j \in \Omega_B} \sum_{n \in \{a,b,c\}} a_{ij}^n Y_{ij}^n = Z_j^n, \;\; i = 1, \ldots, \text{N} \tag{18}$$

The observability function, fi, assigned to bus i, serves as an indicator of its observability, taking into consideration the impact of PZINs. The PZINs must exceed a $3 - \mathit{off}_i^n$ to guarantee the observability of bus i. It is noteworthy that if there is no connection associated with the phase of a bus, observing that specific phase becomes unnecessary. This principle is articulated in Eq. (19).

$$\mathit{off}_i^n \times P_i^n = 0, \quad i = 1, \ldots, \text{N}, \;\; n \in \{a,b,c\} \tag{19}$$

In addition to constraints (16)–(19), two additional constraints are formulated regarding the placement of PZINs in the distribution system to optimize the proposed objective function. This is dictated by the following rule:

1. If a PZIN be connected to a single-end bus, the installation of μ-PMUs is disallowed for all of its zero-injection phases. Conversely, their zero-injection characteristics remain unconstrained. This specification is formally denoted as (20).

$$ZB_j^n XX_j^n = 0, \quad j = 1, \ldots, \text{N}, \;\; \text{n} \in \{a,b,c\} \tag{20}$$

2. If an end node considered as a PZIN, the placement of μ-PMUs is limited for each of its zero-injection phases. This restriction enhances observability by leveraging the PZIN characteristic. These conditions are formally expressed in Eqs. (21) and (22).

$$ZBR_j^n XX_j^n = 0, \; j = 1, \ldots, \mathrm{N}, \; \mathrm{n} \in \{a, b, c\} \tag{21}$$

$$if \; ZBR_j^n = 1, \; Y_{jj}^n = 1 \tag{22}$$

3.2.3 Observability constraints under single μ-PMU outage contingency

In order to depict the occurrence of a single μ-PMU outage, constraint (16) undergoes a modification aimed at improving observability through at least two distinct approaches, detailed in Eq. (23).

$$f_i = \sum_{j \in \Omega_B} \sum_{n \in \{a,b,c\}} \left(\begin{array}{l} a_{ij}^n XX_j^n + \\ 2 \times \left(\sum_{j \in \Omega_B} \sum_{n \in \{a,b,c\}} a_{ij}^n Y_{ij}^n Z_j^n \right) \end{array} \right) \geq 2 \times (3 - off_i^n), i = 1, \ldots, \mathrm{N} \tag{23}$$

The coefficient of 2 in the left part of inequality constraint (23) signifies that if a bus's observability is improved by the PZIN, it maintains observability even in the case of a single μ-PMU outage. This resilience is due to each bus in the system being directly or indirectly observed through a minimum of two distinct pathways. It's crucial to recognize that the radial structure of distribution networks results in some busses being linked to the network via a single line. Although it is theoretically feasible to independently place a μ-PMU at each end bus for multiple observation paths, this is not deemed a cost-effective solution. Consequently, constraint (23) is substituted with (24), where end busses, excluding critical ones, are observed from at least one pathway, while the remaining busses are observed from at least two pathways.

$$f_i = \sum_{j \in \Omega_B} \sum_{n \in \{a,b,c\}} \left(\begin{array}{l} a_{ij}^n XX_j^n + \\ 2 \times \left(\sum_{j \in \Omega_B} \sum_{n \in \{a,b,c\}} a_{ij}^n Y_{ij}^n Z_j^n \right) \end{array} \right) \geq [2 \times (3 - off_i^n)] - R_i^n, i = 1, \ldots, \mathrm{N} \tag{24}$$

3.3 The DSSE problem

The objective of state estimation is to determine accurate values for state variables using real-time measurements obtained from μ-PMUs within the energy management system. The Weighted Least Squares State Estimation is framed as the minimization of the objective function (J(x)), as elaborated in [4].

$$J(x) = \sum_{u=1}^{m} w_u e_u^2 \tag{25}$$

Subjected to:

$$z = h(x) + e \tag{26}$$

3.3.1 Measurement function linearization process

With the exception of voltage magnitudes, most h measurements demonstrate nonlinearity. In this section, Taylor series approximations of sin and cos terms are employed to linearize these nonlinear functions.

$$I_{ij}^{m} = \overline{Y}^{m}.\Delta V_{ij}^{m} \tag{27}$$

Eq. (15) represents the complex power flow S_{ij}^{m} by phase-m.

$$S_{ij}^{m} = V_{i}^{m} \times \left(I_{ij}^{m}\right)^{*} = V_{i}^{m} \times \left(\sum_{m \in \{a,b,c\}} \overline{Y}^{m}.\Delta V_{ij}^{m}\right)^{*} \tag{28}$$

Accordingly, (15) was extended as Eq. (16),

$$S_{ij}^{m} = V_{i}^{m} \measuredangle \delta_{i}^{m}.\left(\sum_{m \in \{a,b,c\}} \left[\left(\overline{G}^{m} + j\overline{B}^{m}\right).\left(\left|V_{i}^{m}\right| \measuredangle \delta_{i}^{m} - \left|V_{j}^{m}\right| \measuredangle \delta_{j}^{m}\right)\right]\right)^{*} \tag{29}$$

Let us denote:

$$\Delta\delta_{ij}^{m} = \delta_{i}^{m} - \delta_{j}^{m} \text{ for } m \in \{a,b,c\} \tag{30}$$

$$\cos_{ij}^{m} = \cos\left(\Delta\delta_{ij}^{m}\right) \text{ and } \sin_{ij}^{m} = \sin\left(\Delta\delta_{ij}^{m}\right) \tag{31}$$

3.3.2 Derivation of non-linear active and reactive power flow equations

$$S_{ij}^{m} = \left|V_{i}^{m}\right| \measuredangle \delta_{i}^{m} \sum_{n \in \{a,b,c\}} \sum_{i,j \in \Omega_B} \begin{bmatrix} \overline{G}^{mn}\left|V_{i}^{n}\right| \measuredangle - \delta_{i}^{n} - \overline{G}^{mn}\left|V_{j}^{n}\right| \measuredangle - \delta_{j}^{n} \\ -j\overline{B}^{mn}\left|V_{i}^{n}\right| \measuredangle - \delta_{i}^{n} + j\overline{B}^{mn}\left|V_{j}^{n}\right| \measuredangle - \delta_{j}^{n} \end{bmatrix} \tag{32}$$

The Eq. (32) was simplified as follows:

$$S_{ij}^{m} = \left|V_{i}^{m}\right| \sum_{n \in \{a,b,c\}} \sum_{i,j \in \Omega_B} \begin{bmatrix} \overline{G}^{mn}\left|V_{i}^{n}\right| \measuredangle \Delta\delta_{ii}^{mn} - \overline{G}^{mn}\left|V_{j}^{n}\right| \measuredangle \Delta\delta_{ij}^{mn} \\ -j\overline{B}^{mn}\left|V_{i}^{n}\right| \measuredangle - \delta_{ii}^{mn} + j\overline{B}^{mn}\left|V_{j}^{n}\right| \measuredangle \Delta\delta_{ij}^{mn} \end{bmatrix} \tag{33}$$

Eq. (33) is separated into its components, distinguishing between its real and imaginary parts, expressed as (34).

$$\begin{aligned} S_{ij}^{m} = {} & \left|V_{i}^{m}\right| \sum_{n \in \{a,b,c\}} \sum_{i,j \in \Omega_B} \left\{\left|V_{i}^{n}\right|\left[\overline{G}^{mn}.\cos_{ii}^{mn} + \overline{B}^{mn}.\sin_{ii}^{mn}\right] - \left|V_{j}^{n}\right|\left[\overline{G}^{mn}.\cos_{ij}^{mn} + \overline{B}^{mn}.\sin_{ij}^{mn}\right]\right\} \\ & + j\left|V_{i}^{m}\right| \sum_{n \in \{a,b,c\}} \sum_{i,j \in \Omega_B} \left\{\left|V_{i}^{n}\right|\left[\overline{G}^{mn}.\sin_{ii}^{mn} - \overline{B}^{mn}.\cos_{ii}^{mn}\right] - \left|V_{j}^{n}\right|\left[\overline{G}^{mn}.\sin_{ij}^{mn} + \overline{B}^{mn}.\cos_{ij}^{mn}\right]\right\} \end{aligned} \tag{34}$$

The actual segment of Eq. (34) outlines the equation for active power flow concerning phase *m* and line *ij*, as follows:

$$P_{ij}^{m} = |V_i^m| \sum_{n \in \{a,b,c\}} \sum_{i,j \in \Omega_B} \left\{ \begin{array}{l} |V_i^n| \left[\overline{G}^{mn} . \cos_{ii}^{mn} + \overline{B}^{mn} . \sin_{ii}^{mn}\right] - \\ \left|V_j^n\right| \left[\overline{G}^{mn} . \cos_{ij}^{mn} + \overline{B}^{mn} . \sin_{ij}^{mn}\right] \end{array} \right\} \tag{35}$$

The imaginary component of Eq. (34) delineates the equation for reactive power flow related to phase m and line ij, as follows:

$$Q_{ij}^{m} = |V_i^m| \sum_{n \in \{a,b,c\}} \sum_{i,j \in \Omega_B} \left\{ \begin{array}{l} |V_i^n| \left[\overline{G}^{mn} . \sin_{ii}^{mn} - \overline{B}^{mn} . \cos_{ii}^{mn}\right] \\ -\left|V_j^n\right| \left[\overline{G}^{mn} . \sin_{ij}^{mn} + \overline{B}^{mn} . \cos_{ij}^{mn}\right] \end{array} \right\} \tag{36}$$

Eqs. (37)–(40) represent the outcomes of the Taylor series approximations for sin and cos functions

$$\sin_{ii}^{mn} = \sin(\Delta\delta_{ii}^{mn}) \approx \sin(\Delta\delta_{ii}^{mn,r}) + \cos(\Delta\delta_{ii}^{mn,r}).\Delta\delta_{ii}^{mn} \tag{37}$$

$$\cos_{ii}^{mn} = \cos(\Delta\delta_{ii}^{mn}) \approx \cos(\Delta\delta_{ii}^{mn,r}) - \sin(\Delta\delta_{ii}^{mn,r}).\Delta\delta_{ii}^{mn} \tag{38}$$

$$\sin_{ij}^{mn} = \sin\left(\Delta\delta_{ij}^{mn}\right) \approx \sin\left(\Delta\delta_{ij}^{mn,r}\right) + \cos\left(\Delta\delta_{ij}^{mn,r}\right).\Delta\delta_{ij}^{mn} \tag{39}$$

$$\cos_{ij}^{mn} = \cos\left(\Delta\delta_{ij}^{mn}\right) \approx \cos\left(\Delta\delta_{ij}^{mn,r}\right) - \sin\left(\Delta\delta_{ij}^{mn,r}\right).\Delta\delta_{ij}^{mn} \tag{40}$$

The operational points for voltage magnitudes were determined regionally for each line section by utilizing the combined data from μ-PMUs, aiming to improve the accuracy of state estimation. In the context of a specific line section denoted as ij, with q representing the count of μ-PMU voltage measurements, the local operating point of the voltage magnitude, denoted as $|V_r|$, is calculated using Eq. (41).

$$|V_r| \approx \frac{1}{q} \sum_{i=1}^{q} |V_i| \tag{41}$$

3.3.3 Derivation of linearized active and reactive power flow equations

In Eq. (35), the nonlinear active power Eq. (42) for phase m is depicted. To linearize the nonlinear sin(.) and cos(.) equations, they have been replaced with Taylor series approximations (37)–(40), as detailed below:

$$\overline{P}_{ij}^{m} = |V_i^m| \sum_{n \in \{a.b,c\}} \sum_{i,j \in \Omega_B} \left\{ \begin{array}{l} |V_i^n| \left[\begin{array}{l} \overline{G}^{mn}.\left(\cos(\Delta\delta_{ii}^{mn,r}) - \sin(\Delta\delta_{ii}^{mn,r}).\Delta\delta_{ii}^{mn}\right) \\ +\overline{B}^{mn}.\left(\sin(\Delta\delta_{ii}^{mn,r}) + \cos(\Delta\delta_{ii}^{mn,r}).\Delta\delta_{ii}^{mn}\right) \end{array} \right] \\ -\left|V_j^n\right| \left[\begin{array}{l} \overline{G}^{mn}.\left(\cos\left(\Delta\delta_{ij}^{mn,r}\right) - \sin\left(\Delta\delta_{ij}^{mn,r}\right).\Delta\delta_{ij}^{mn,r}\right) \\ +\overline{B}^{mn}.\left(\sin\left(\Delta\delta_{ij}^{mn,r}\right) + \cos\left(\Delta\delta_{ij}^{mn,r}\right).\Delta\delta_{ij}^{mn}\right) \end{array} \right] \end{array} \right\} \tag{42}$$

The assumption $\Delta\delta_{ii}^{mn,r} \approx \Delta\delta_{ij}^{mn,r}$ is made that the distance between two adjacent busses in the distribution network is minimal. Based on this assumption, Eq. (42) is revised as (43):

$$\overline{P}_{ij}^{m} = |V_i^m| \sum_{n\in\{a.b,c\}} \sum_{i,j\in\Omega_B} \left\{ \begin{array}{l} |V_i^n| \left[\begin{array}{l} \overline{G}^{mn}.\left(\cos\left(\Delta\delta_{ij}^{mn,r}\right) - \sin\left(\Delta\delta_{ij}^{mn,r}\right).\Delta\delta_{ii}^{mn}\right) \\ +\overline{B}^{mn}.\left(\sin\left(\Delta\delta_{ij}^{mn,r}\right) + \cos\left(\Delta\delta_{ij}^{mn,r}\right).\Delta\delta_{ii}^{mn}\right) \end{array} \right] \\ -\left|V_j^n\right| \left[\begin{array}{l} \overline{G}^{mn}.\left(\cos\left(\Delta\delta_{ij}^{mn,r}\right) - \sin\left(\Delta\delta_{ij}^{mn,r}\right).\Delta\delta_{ij}^{mn,r}\right) \\ +\overline{B}^{mn}.\left(\sin\left(\Delta\delta_{ij}^{mn,r}\right) + \cos\left(\Delta\delta_{ij}^{mn,r}\right).\Delta\delta_{ij}^{mn}\right) \end{array} \right] \end{array} \right\} \tag{43}$$

Eq. (43) was rearranged as:

$$\overline{P}_{ij}^{m} = |V_i^m| \sum_{n\in\{a,b,c\}} \sum_{i,j\in\Omega_B} \left\{ \begin{array}{l} \left(\overline{G}^{mn}.\cos\left(\Delta\delta_{ij}^{mn,r}\right) + \overline{B}^{mn}.\sin\left(\Delta\delta_{ij}^{mn,r}\right)\right).\left(|V_i^n| - \left|V_j^n\right|\right) \\ +\left(-\overline{G}^{mn}.\sin\left(\Delta\delta_{ij}^{mn,r}\right) + \overline{B}^{mn}.\cos\left(\Delta\delta_{ij}^{mn,r}\right)\right).\left(|V_i^n|.\Delta\delta_{ii}^{mn} - \left|V_j^n\right|.\Delta\delta_{ij}^{mn}\right) \end{array} \right\} \tag{44}$$

The simplification of the regional operating points for voltage magnitudes is expressed by Eq. (44).

$$\overline{P}_{ij}^{m} = \sum_{n\in\{a,b,c\}} \sum_{i,j\in\Omega_B} \left\{ \begin{array}{l} |V_r|\left(\overline{G}^{mn}.\cos\left(\Delta\delta_{ij}^{mn,r}\right) + \overline{B}^{mn}.\sin\left(\Delta\delta_{ij}^{mn,r}\right)\right).\left(|V_i^n| - \left|V_j^n\right|\right) \\ +\left|V_r^2\right|\left(-\overline{G}^{mn}.\sin\left(\Delta\delta_{ij}^{mn,r}\right) + \overline{B}^{mn}.\cos\left(\Delta\delta_{ij}^{mn,r}\right)\right).\left(\Delta\delta_{ii}^{mn} - \Delta\delta_{ij}^{mn}\right) \end{array} \right\} \tag{45}$$

As earlier described in (30) and (31), it is clear that $\Delta\delta_{ii}^{mn} = \delta_i^m - \delta_i^n$ and $\Delta\delta_{ij}^{mn} = \delta_i^m - \delta_j^n$. Therefore, $\Delta\delta_{ii}^{mn} - \Delta\delta_{ij}^{mn} = -\left(\delta_i^n - \delta_j^n\right)$. The linearized active power function for phase-m $\overline{P}_{ij}^{m}$ was constructed by replacing this for (46), as follows:

$$\overline{P}_{ij}^{m} = \sum_{n\in\{a,b,c\}} \sum_{i,j\in\Omega_B} \left\{ \begin{array}{l} |V_r|\left(\overline{G}^{mn}.\cos\left(\Delta\delta_{ij}^{mn,r}\right) + \overline{B}^{mn}.\sin\left(\Delta\delta_{ij}^{mn,r}\right)\right).\left(|V_i^n| - \left|V_j^n\right|\right) \\ -\left|V_r^2\right|\left(-\overline{G}^{mn}.\sin\left(\Delta\delta_{ij}^{mn,r}\right) + \overline{B}^{mn}.\cos\left(\Delta\delta_{ij}^{mn,r}\right)\right).\left(\delta_i^n - \delta_j^n\right) \end{array} \right\} \tag{46}$$

Eq. (47) represents the result of linearizing the reactive power function $\overline{Q}_{ij}^{m}$ for phase-m.

$$\overline{Q}_{ij}^{m} = \sum_{n\in\{a,b,c\}} \sum_{i,j\in\Omega_B} \left\{ \begin{array}{l} |V_r|\left(\overline{G}^{mn}.\sin\left(\Delta\delta_{ij}^{mn,r}\right) - \overline{B}^{mn}.\cos\left(\Delta\delta_{ij}^{mn,r}\right)\right).\left(|V_i^n| - \left|V_j^n\right|\right) \\ -\left|V_r^2\right|\left(-\overline{G}^{mn}.\cos\left(\Delta\delta_{ij}^{mn,r}\right) + \overline{B}^{mn}.\sin\left(\Delta\delta_{ij}^{mn,r}\right)\right).\left(\delta_i^n - \delta_j^n\right) \end{array} \right\} \tag{47}$$

With reference to the voltage magnitude operating points defined in Eq. (41) and employing Taylor series approximations on (37)–(40), the refined linearized approximations for the active and reactive power equations were revealed as,

$$\overline{P}_{ij}^{m} = \sum_{n \in \{a,b,c\}} \sum_{i,j \in \Omega_B} \left\{ |V_r|.\Phi_P.\left(|V_i^n| - |V_j^n| \right) - |V_r^2|.\Psi_P.\left(\delta_i^n - \delta_j^n \right) \right\} \tag{48}$$

$$\overline{Q}_{ij}^{m} = \sum_{n \in \{a,b,c\}} \sum_{i,j \in \Omega_B} \left\{ |V_r|.\Phi_Q.\left(|V_i^n| - |V_j^n| \right) - |V_r^2|.\Psi_Q.\left(\delta_i^n - \delta_j^n \right) \right\} \tag{49}$$

Following the linearization procedure, the measurement function presented in (1) was reformulated as a linear function (50).

$$z = H.x + e \tag{50}$$

Utilizing the WLS method, the computed estimated state vector was determined according to Eq. (51).

$$\hat{x} = \left[\left(H^T U^{-1} H \right)^{-1} H^T U^{-1} \right] z \tag{51}$$

Eq. (52) illustrates the diagonal matrix that includes the respective measurement variances.

$$U = \text{diag}\left(\sigma_1^2, \sigma_2^2, \sigma_3^2, \ldots, \sigma_m^2 \right) \tag{52}$$

Eq. (53) demonstrates the measurement vector, encompassing voltage magnitude, voltage angle, as well as active and reactive power flows for each phase.

$$Z = \left[|V_i^a|, |V_i^b|, |V_i^c|, \delta_i^a, \delta_i^b, \delta_i^c, P^a, Q^a, P^b, Q^b, P^c, Q^c \right]^T \tag{53}$$

4. Two-stage stochastic programming

Addressing decision-making challenges in an uncertain setting involves the introduction of an innovative stochastic two-stage programming model designed to optimize both μ-PMU placement and DSSE problems. This methodology entails the classification of decision variables into two main categories: "*here-and-now*" and "*wait-and-see*," as detailed in Refs. [34, 35]. In this methodology, the first phase entails establishing the real values of bus voltage and branch current phasors while guaranteeing the observability of the system. The outcomes from the initial stage subsequently act as input for the succeeding stage. The first stage involves dual variables identified as '*d*' (representing voltage phasors) and variables '*c*' (representing current phasors), necessitating immediate decisions in the present, before uncertainties are resolved. Simultaneously, operational variables '*xt*' function as components for a "*wait-and-see*" approach, and the determination of variable '*b*' in the second stage can be made once all uncertain parameters have been observed. The formulated stochastic two-stage optimization is presented as follows:

$$\begin{aligned} (\text{First\_stage}): \quad & \max_{c,d} \; F_f(c,d) + E\left[F_f(c,d,\xi) \right] \\ \text{S.t.} \quad & \varphi_f(c,d) = 0 \\ & \psi_f(c,d) \leq 0 \\ & c \in \Re^1 \end{aligned} \tag{54}$$

$$
\begin{aligned}
(\text{Second\_stage}) : F_s(c,d,\xi) &= \min_{x_t} \; F_s(b,c,d,x_t,\xi) \\
\text{S.t.} \; \varphi_s(b,c,d,x_t,\xi) &= 0 \\
\psi_s(b,c,d,x_t,\xi) &\le 0 \\
b \in \{0,1\}^m, x_t &\in \Re^n
\end{aligned}
\tag{55}
$$

Where, the objective function is separated into two parts: a deterministic element denoted as F_f, which reflects decisions, and the expected value of a stochastic element, F_s. The stochastic component is influenced by the actualization of uncertain parameters 'n' during the operational stage.

5. Simulation results and discussion

This chapter introduces a two-stage stochastic programming model designed for asymmetrical network state estimation, aiming to enhance SA through the incorporation of μ-PMU data. **Figure 1** visually outlines the proposed model. In the initial stage, a dynamic and efficient mathematical programming challenge, as outlined in Ref. [36], harnesses the capabilities of the CPLEX optimizer to address the optimal placement of μ-PMUs amidst the presence of PZINs and various contingencies. The use of this optimizer streamlines decision optimization, leading to enhanced efficiency, cost reduction, and increased profitability. Its proficiency in handling large-scale, real-world problems aligns well with the swift processing requirements of contemporary interactive decision optimization applications. Moreover, in the second phase, the linear approximation of active and reactive power functions is employed to represent

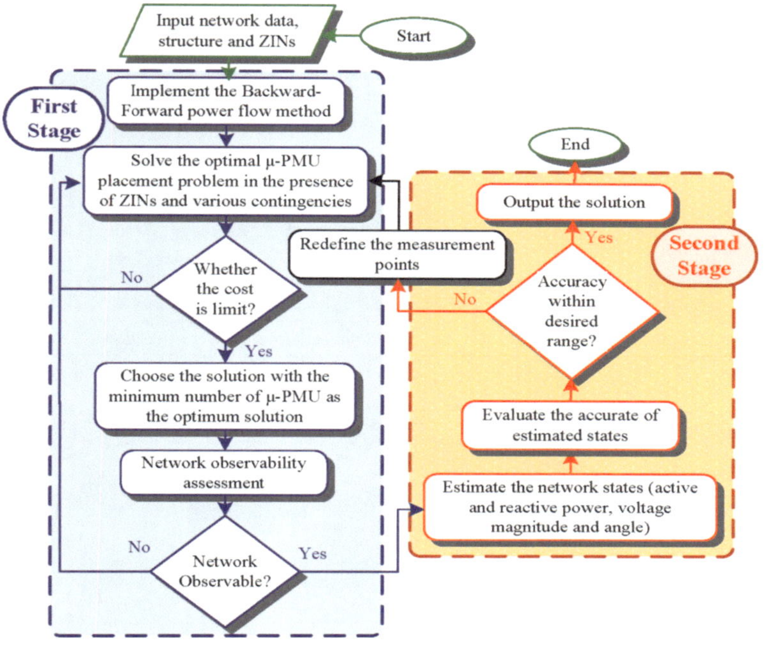

Figure 1.
The flowchart of the proposed two-stage programming problem.

the linear DSSE. Finally, the Weighted Least Squares algorithm is utilized to address the DSSE problem integrated with three-phase DG. The resolution of the proposed two-stage programming problem is conducted within the MATLAB environment, as elaborated in Ref. [37].

This chapter investigates four scenarios for the operation and analysis of the distribution system under different operational conditions, allowing for a comparative assessment of the outcomes.

- **Scenario I:** Normal operational circumstances
- **Scenario II:** Normal operational circumstances with consideration of PZINs
- **Scenario III:** Contingency operational conditions
- **Scenario IV:** Contingency operational conditions with consideration of PZINs

5.1 Case study

To offer a thorough understanding of asymmetrical distribution networks, we have opted for the modified IEEE 85-bus test feeder, as presented in Ref. [38]. The single-line diagram of this system is illustrated in **Figure 2**, comprising 85 busses and 84 branches, operating at 11 kV with a 100 MVA base. The phases are indicated by red, blue, and green lines corresponding to a, b, and c phases, respectively. Furthermore, the system incorporates nine Distributed Generation sources, including wind turbines, solar panels, and combined heat and power systems, connected to various busses. **Table 1** provides technical details for these sources. It is important to note that all μ-PMUs are expected to be of the same type, ensuring uniform costs. The cost estimation for each μ-PMU device, including installation, is $1000 across all test systems, as detailed in the reference [40].

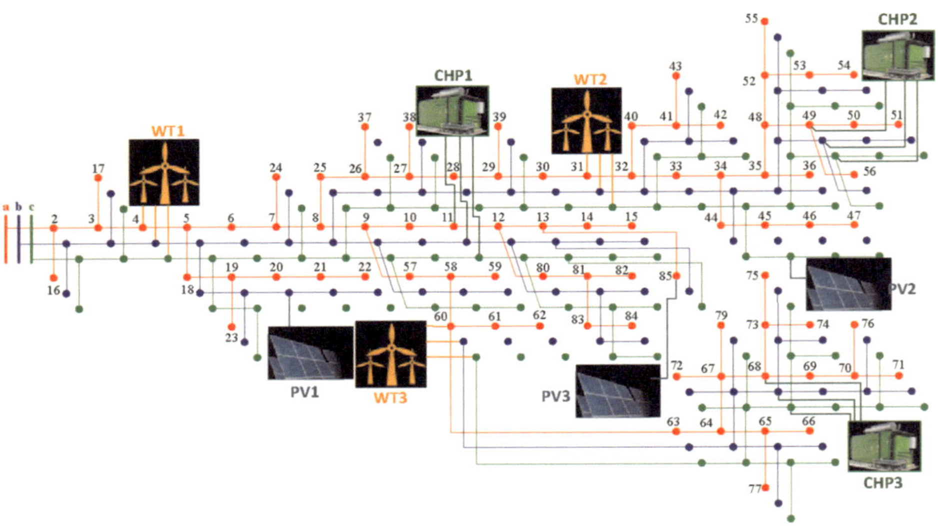

Figure 2.
A modified asymmetric three-phase 85-bus distribution test system.

DG type	Location (bus)	Lower bound (kW)	Upper bound (kW)	Cost coefficient ($/kWh)
WT1	4	200	450	0.02
WT2	31	200	450	0.02
WT3	60	100	360	0.02
PV1	20	150	350	0.05
PV2	45	150	350	0.05
PV3	85	150	350	0.05
CHP1	28	300	700	0.01
CHP2	49	350	750	0.01
CHP3	68	300	700	0.01

Table 1.
Technical data of utilized DG units [39].

5.1.1 Scenario I

Determine optimal locations and quantities of single-phase and three-phase μ-PMUs, along with their associated redundancy indices and installation costs, to achieve complete observability in asymmetrical distribution networks during normal operational conditions. This evaluation, excluding considerations for partially zero injection nodes, is concisely summarized in the initial row of **Table 2**.

5.1.2 Scenario II

In the prior analysis, the emphasis was on determining the most suitable allocation of μ-PMUs in typical operational scenarios, excluding considerations of PZINs. The results of investigating the optimal μ-PMU allocation under standard operational conditions, now accounting for PZINs, are outlined in the second row of **Table 2**. This underscores the effectiveness of the proposed methodology. Noteworthy is the observation that integrating PZINs into the model leads to a reduction in the required number of μ-PMUs and a lowering of the objective function value. The findings underscore that a decrease in the number of μ-PMUs corresponds to a reduction in the measurement redundancy index, ultimately enhancing measurement efficiency.

5.1.3 Scenario III

The third row of **Table 2** displays the optimal locations and quantities of μ-PMUs, installation costs, and the resulting measurement redundancy index, ensuring full observability of the system under various contingencies, without considering PZINs. When a μ-PMU is placed at a specific node, it not only enhances visibility at that bus but also indirectly contributes to the visibility of neighboring busses. Consequently, in the event of a single line outage or μ-PMU failure, certain busses may lose observability. To address this issue, an optimal μ-PMU allocation approach is employed to identify the best locations for maintaining complete system observability after any contingency. It is evident that, with the occurrence of one or more events, the number

Network name	Contingency statue	PZINs locations	Optimal places of μ-PMUs	No. of μ-PMUs	Measurement Redundancy	Obj. Fun. value	Exe. time (s)
Modified asymmetric three-phase 85 bus distribution test system	Under normal operating condition	—	2, 3, 5, 7, 13, 14, 19, 27, 29, 32, 34, 35, 41, 49, 50, 52, 53, 58, 60, 67, 70, 73, 81, 83: All three phases 21: Phases a & b 26, 46: Phases a & c 10, 61, 65: Phase a	Three-phase units: 27 Single-phase units: 3	114	27.0637	1.1075
	Under normal operating condition	2, 3, 5, 7, 9, 10, 12, 13, 27, 29, 32, 34, 35, 41, 48, 49, 52, 58, 64, 65, 67, 68, 70, 73, 81	2, 4, 7, 12, 14, 19, 26, 29, 33, 41, 50, 53, 57, 64, 65, 67, 70, 73, 83: All three phases 21: Phases a & b 46: Phases a & c 61: Phase a	Three-phase units: 21 Single-phase units: 1	79	19.2640	1.5553
	Single line outage or single μ-PMU loss	—	2, 3, 5, 6, 7, 8, 9, 10, 12, 13, 14, 15, 17, 19, 23, 24, 26, 27, 29, 31, 32, 34, 35, 36, 38, 39, 40, 41, 49, 50, 51, 52, 53, 54, 55, 56, 58, 59, 60, 64, 65, 67, 68, 69, 70, 73, 81, 82, 83, 84: All three phase 21, 22: Phases a & b 37, 45, 46, 47: Phases a & c 61, 62, 78, 85: Phase a	Three-phase units: 56 Single-phase units: 4	189	58.1235	1.3018
	Single line outage or single μ-PMU loss	2, 3, 5, 7, 9, 10, 12, 13, 27, 29, 32, 34, 35, 41, 48, 49, 52, 58, 64, 65, 67, 68, 70, 73, 81	1, 2, 3, 5, 7, 8, 9, 11, 14, 15, 19, 23, 26, 28, 29, 32, 34, 40, 41, 44, 48, 50, 51, 53, 54, 58, 64, 65, 66, 67, 68, 69, 70, 73, 75, 79, 81, 83, 84 21, 22: Phases a & b 37, 46, 47: Phases a & c 61, 62: Phase a	Three-phase units: 44 Single-phase units: 2	147	42.8835	1.7802

Table 2.
Results of optimal allocation in a combination of single and three-phase μ-PMUs with and without considering PZINs under contingencies.

of installed μ-PMUs in the system increases, leading to a significant rise in the objective function value. Consequently, there is no economic justification for scenario III.

5.1.4 Scenario IV

Table 2 provides a detailed breakdown of the optimal setup involving μ-PMUs, installation costs, and the measurement redundancy index for the proposed distribution network. This comprehensive information is specifically located in the fourth row of the table and is examined across various contingencies, ensuring the network's complete observability following the incorporation of PZINs. As anticipated, additional μ-PMUs are deemed necessary to ensure observability under diverse contingencies. The viability of scenario III hinges on the essential integration of PZINs to minimize the objective function, leading to a simultaneous reduction in both the required number of μ-PMUs and the overall system objective function value. The incorporation of PZINs into the allocation problem proves to be a pivotal factor, resulting in a noteworthy decrease in the essential μ-PMUs and the corresponding system objective function value.

In this segment, we present the outcomes of the three-phase state estimation, where the estimated values are compared with the actual values derived from the initial power flow computations for each specific scenario. The data input for the state estimator comprises a dataset featuring active and reactive power flow, voltage magnitude and angle, in addition to DGs input. This dataset is constructed using a series of voltage and current phasors generated by μ-PMUs. The provided passage delves into the outcomes of the proposed methodology aimed at estimating active power flow within a three-phase 85-bus distribution network. In **Figure 3**, a visual representation showcases the comparison between estimated and actual values across phases (a, b, and c). The maximum variances in active power flow estimation stand at 0.0716, 0.0918, and 0.1799 per unit for lines 17, 17, and 7, respectively. Similarly, **Figure 4** exhibits the actual and estimated values for three-phase reactive powers at designated load points. The highest variances in reactive power flow estimation in phases a, b,

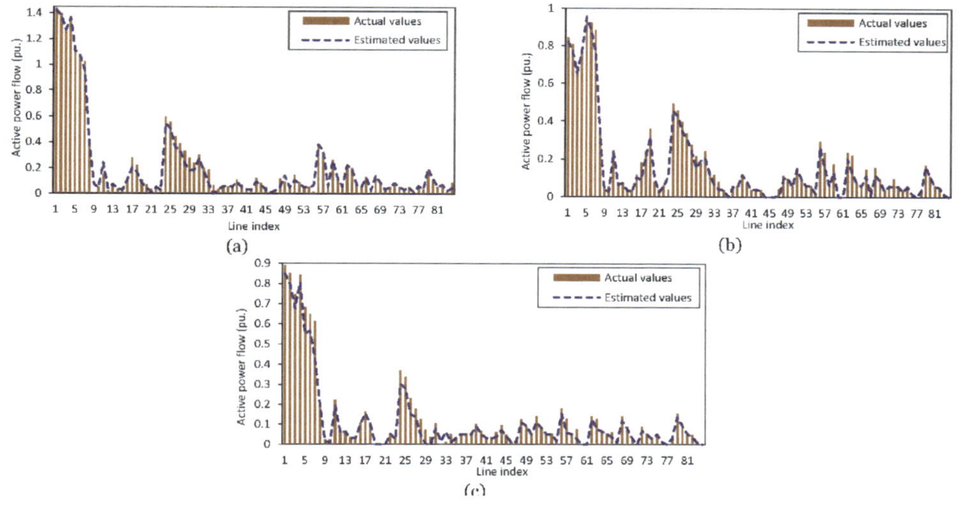

Figure 3.
Estimated and actual values of the active power flow.

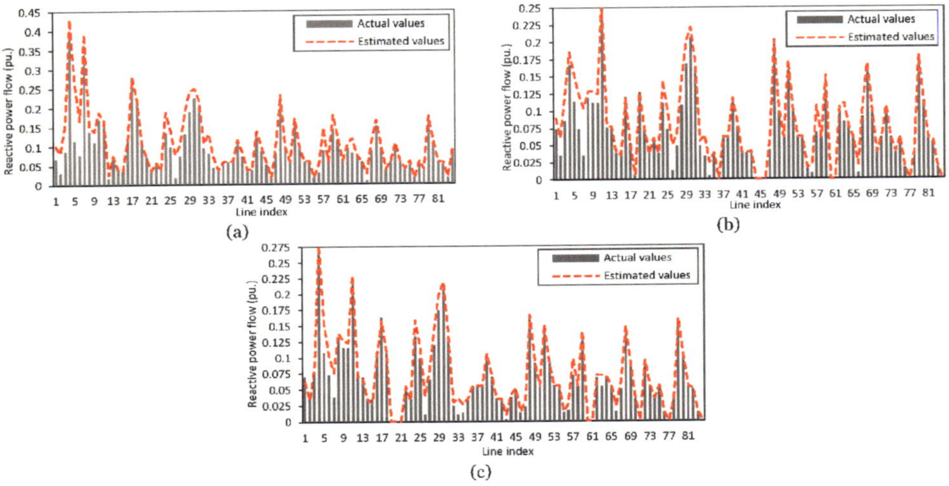

Figure 4.
Estimated and actual values of the reactive power flow.

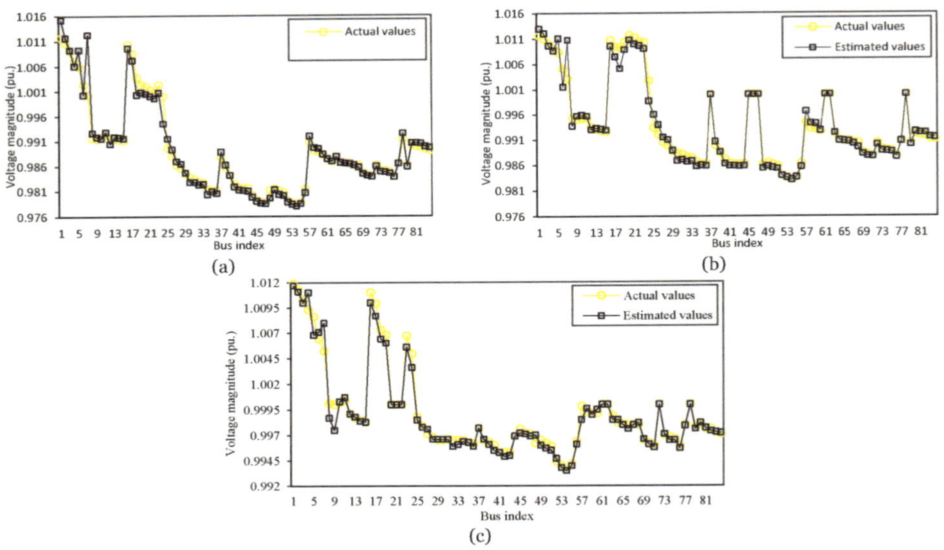

Figure 5.
Estimated and actual values of voltage magnitudes.

and c are 0.1477, 0.0697, and 0.0387 per unit, associated with lines 5, 7, and 7, respectively. **Figures 2** and **3** underscore the remarkable proximity between estimated and actual values, providing compelling evidence for the efficacy of the two-stage programming approach for DSSE.

In **Figure 5**, the voltage magnitudes for the three phases in the 85-bus model are presented, comparing the estimated values to the actual measurements. In cases where phases are not present, a placeholder value of one has been employed. Visual inspection indicates a substantial agreement between the estimated and actual voltage magnitudes. Similarly, **Figure 6** displays the estimated and actual voltage angles,

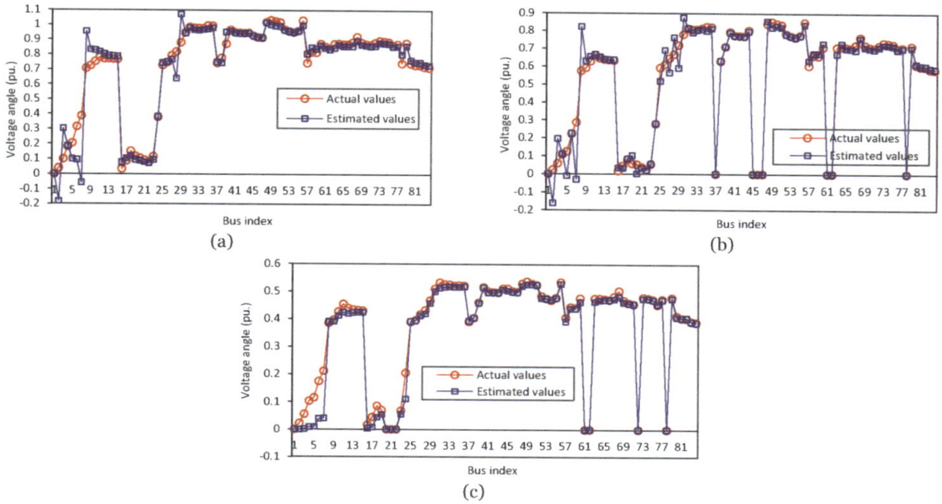

Figure 6.
Estimated and actual values of voltage angles.

assuming a value of zero for the phases that are not present. The proposed methodology consistently achieves heightened accuracy in estimating these parameters. As demonstrated in the presentation, the state estimation model outlined in this study successfully converges and accurately determines the state variables in the adapted asymmetric three-phase 85-bus distribution test system, which incorporates DGs. The simulation results affirm the effectiveness of both the suggested state estimation model and the extended per-unit system. **Figure 7** illustrates the estimated and actual values of DG production for all DG types. The most notable disparity in DG production estimation is observed at 0.0265 per unit, specifically associated with PV_3.

In the proposed DSSE approach, the variances for active and reactive power flows are set at 5×10^{-4} and 2×10^{-4}, respectively. Similarly, the variances governing voltage magnitude and voltage angle are adjusted to 2×10^{-4} and 6×10^{-4}, respectively. Over the course of this investigation, the solution variables undergo a total of 17 updates. During each update, the freshly revised solutions are systematically juxtaposed with their predecessors, and only those demonstrating superior accuracy are chosen as the definitive solutions. **Figures 8** and **9** illustrate the forecasted errors in estimating active and reactive power flows. Significantly, the proposed method demonstrates improved accuracy, with differences between estimated and actual values leading to errors in voltage magnitude and phase angle, benchmarked against load

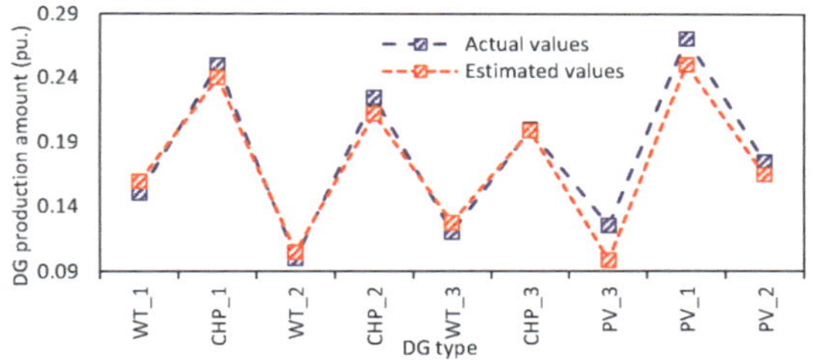

Figure 7.
Estimated and actual values of DG production amount.

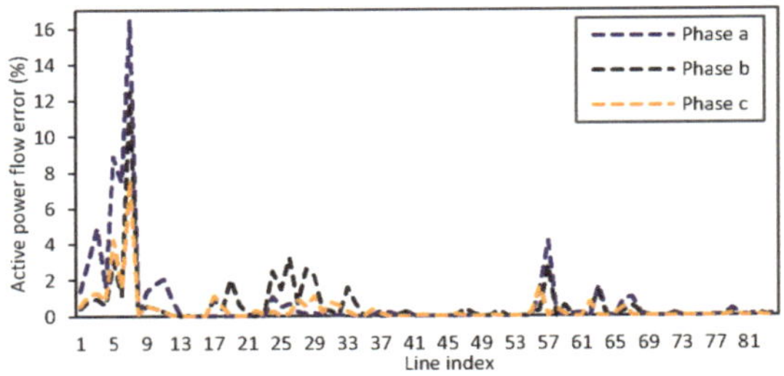

Figure 8.
Active power flow error.

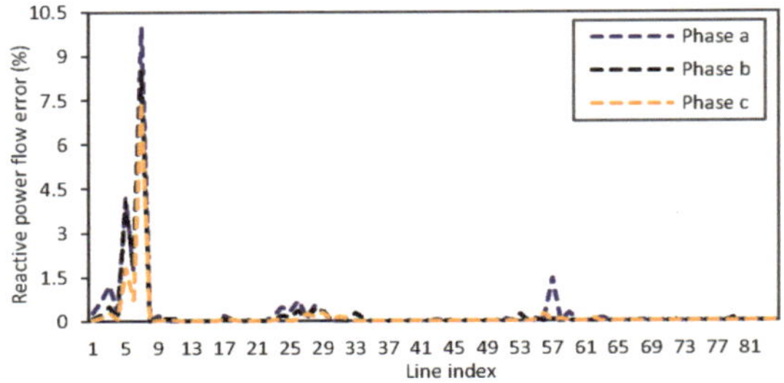

Figure 9.
Reactive power flow error.

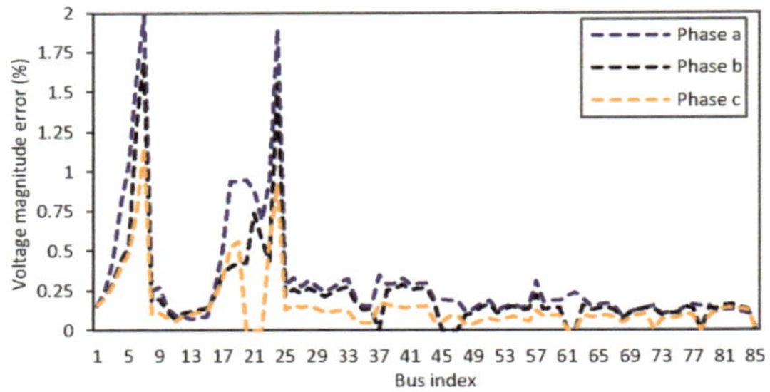

Figure 10.
Voltage magnitude error.

flow results. In **Figure 10**, the errors in voltage magnitude for all nodes are presented, obtained through the proposed state estimation after addressing the μ-PMU placement problem in the initial stage. More precisely, our observations reveal that the proposed technique achieves markedly heightened accuracy when addressing this specific challenge. In **Figure 11**, we depict the expected errors in estimation, concentrating specifically on the estimated voltage phase angle at each node within the 85-bus distribution system. This illustration highlights variations across three separate phases. Recognizing the crucial significance of phase angle measurement in power system state estimation—where even slight angle fluctuations can trigger system instability—the presented results underscore the efficacy of the suggested technique. Following the strategic placement of optimal μ-PMUs at designated busses, the state estimation reveals minimal deviation in angle errors. In order to highlight the effectiveness and merits of the proposed model in comparison to previous studies, diverse analyses have been undertaken.

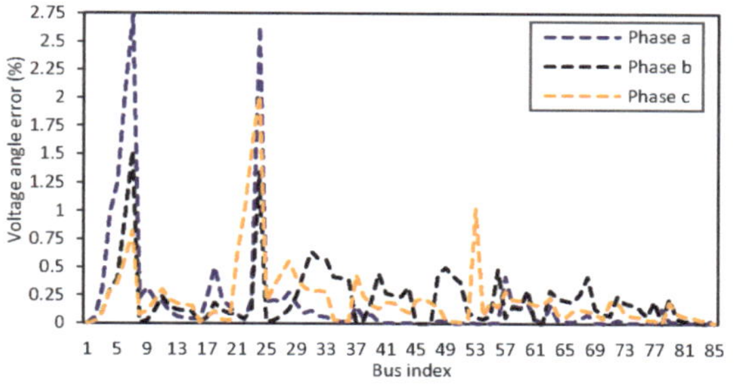

Figure 11.
Voltage angle error.

The proposed method for distribution system state estimation utilizes the EKF approach in addition to the weighted least square (WLS) method. The results of the state estimation using the EKF approach are presented in **Table 3**. The actual values are obtained from the backward-forward power flow method, while the estimated values are obtained from the EKF approach. Comparing the actual and estimated

Voltage magnitude		Voltage angle		Active power	
Actual value	**Estimated value**	**Actual value**	**Estimated value**	**Actual value**	**Estimated value**
1.0117	1.0112	0	0	1.4343	1.4328
1.0106	1.0101	0.0394	−0.162	1.3975	1.3909
1.009	1.0085	0.099	0.2828	1.2832	1.2815
1.0071	1.0066	0.1818	0.1882	1.3748	1.3696
1.0059	1.0054	0.2052	0.1993	1.0979	1.1028
1.0023	1.0018	0.3155	0.394	1.0589	1.0599
1.0002	0.9997	0.386	−0.0981	1.0214	0.9834
0.9915	0.991	0.7059	0.6558	0.3817	0.4087
0.9913	0.9908	0.7235	0.9318	0.0051	0.0025
0.9916	0.9911	0.7512	0.9277	0.051	0.0321
0.9922	0.9917	0.7805	0.8379	0.1429	0.2705
0.9913	0.9908	0.7693	0.7934	0.019	0.0583
0.9914	0.9909	0.77	0.798	0.0706	0.0789
0.9912	0.9907	0.7674	0.7984	0.0353	0.4900
0.991	0.9905	0.7658	0.7966	0.0353	0.0819
1.0103	1.0133	0.0333	0.0681	0.1121	0.1070
1.0085	1.0115	0.0869	0.0949	0.2754	0.1960
1.0037	1.0067	0.1512	0.1412	0.2185	0.1650
1.0023	1.0053	0.1178	0.1047	0.1061	0.0763
1.0019	1.0049	0.1062	0.1082	0.0707	0.0401
1.0012	1.0042	0.0922	0.0893	0.0353	0.034
1.0007	1.0037	0.079	0.0817	0.056	0.0432

Voltage magnitude		Voltage angle		Active power	
Actual value	**Estimated value**	**Actual value**	**Estimated value**	**Actual value**	**Estimated value**
1.0022	1.0052	0.1153	0.1045	0.0353	0.0458
0.9999	0.9997	0.3782	0.3943	0.5958	0.6099
0.9895	0.9893	0.7269	0.7336	0.5592	0.5892
0.9879	0.9877	0.7468	0.7607	0.4464	0.3546
0.9861	0.9859	0.792	0.7752	0.3896	0.3329
0.9855	0.9853	0.8185	0.7587	0.3333	0.2399
0.9844	0.9842	0.8791	0.9708	0.2767	0.191
0.9837	0.9835	0.9476	0.9529	0.241	0.1799
0.9833	0.9853	0.9843	0.9691	0.3055	0.259
0.9828	0.9848	0.9785	0.9542	0.1992	0.1734
0.9826	0.9846	0.9776	0.9731	0.1852	0.0642
0.9813	0.9833	0.9767	0.9769	0.0647	0.0547
0.9811	0.9831	0.9952	0.9612	0.0353	0.0228
0.981	0.983	0.9936	0.9689	0.056	0.597
0.9877	0.9897	0.7418	0.7508	0.056	0.0349
0.9856	0.9876	0.778	0.7337	0.056	0.0441
0.9841	0.9861	0.8715	0.851	0.1061	0.0949
0.9824	0.9844	0.9664	0.9679	0.0707	0.0426
0.9816	0.9836	0.9487	0.9452	0.0353	0.0449
0.9815	0.9835	0.9462	0.9484	0.0353	0.0351
0.9815	0.9835	0.9446	0.9622	0.1202	0.0851
0.9801	0.9821	0.9464	0.9526	0.0847	0.0682
0.9792	0.9812	0.9269	0.9287	0.0493	0.0350
0.9788	0.9808	0.9156	0.9152	0.014	0.0102
0.9787	0.9807	0.9136	0.9145	0.0293	0.0241
0.9811	0.9831	1.0191	1.0171	0.1185	0.0865
0.9814	0.9834	1.0328	1.0218	0.0924	0.1484
0.981	0.983	1.0244	1.0933	0.056	0.0560
0.9807	0.9827	1.018	1.008	0.1478	0.1176
0.979	0.981	0.9684	0.9649	0.0914	0.803
0.9786	0.9806	0.9579	0.9589	0.056	0.0515
0.9782	0.9782	0.9501	0.9458	0.056	0.0573
0.9787	0.9787	0.9606	0.9581	0.014	0.0222
0.9813	0.9813	1.0309	0.998	0.3765	0.0383
0.9906	0.9906	0.744	0.8056	0.3202	0.3461
0.9889	0.9889	0.8173	0.848	0.056	0.0573
0.9888	0.9888	0.8147	0.8397	0.2635	0.2590

Voltage magnitude		Voltage angle		Active power	
Actual value	**Estimated value**	**Actual value**	**Estimated value**	**Actual value**	**Estimated value**
0.9882	0.9882	0.8737	0.8609	0.1122	0.1160
0.9873	0.9873	0.8534	0.8427	0.056	0.0610
0.9868	0.9868	0.8395	0.8329	0.215	0.2248
0.9878	0.9878	0.8733	0.8455	0.2009	0.2309
0.9866	0.9866	0.8747	0.8649	0.07	0.0710
0.9865	0.9865	0.8715	0.8563	0.056	0.0562
0.9864	0.9864	0.8689	0.8547	0.1307	0.1316
0.9861	0.9861	0.8835	0.8567	0.0393	0.0312
0.9861	0.9861	0.9217	0.8926	0.1477	0.1431
0.9846	0.9846	0.8816	0.8657	0.0914	0.0814
0.9841	0.9841	0.8712	0.8594	0.0353	0.0351
0.9839	0.9839	0.8664	0.8562	0.056	0.036
0.9861	0.9861	0.8809	0.8593	0.0915	0.0851
0.985	0.985	0.8947	0.878	0.056	0.0554
0.985	0.9846	0.8909	0.8755	0.0353	0.0345
0.9847	0.9843	0.8859	0.8735	0.056	0.0555
0.9838	0.9834	0.8635	0.8538	0.014	0.0141
0.9865	0.9861	0.8712	0.8565	0.056	0.0561
0.9912	0.9908	0.7424	0.8023	0.0353	0.0350
0.986	0.9856	0.8786	0.8584	0.1617	0.2101
0.9901	0.9897	0.7403	0.7621	0.1054	0.1317
0.9897	0.9893	0.7308	0.7509	0.056	0.059
0.9896	0.9892	0.7295	0.7494	0.0493	0.0632
0.9892	0.9888	0.7174	0.7341	0.014	0.0150
0.989	0.9886	0.714	0.7286	0.0896	0.0410
0.9921	0.9917	0.787	0.7943		

Table 3.
Results of distribution system states obtained from EKF method.

values, it can be observed that the estimated values closely approximate the actual values. This demonstrates that the proposed state estimation method is highly efficient and robust, with minimal errors. Numerical analysis reveals that the EKF approach is slightly more accurate than WLS. However, it should be noted that the EKF utilizes both past and present measurements, whereas WLS only considers measurements from the current time-step. Therefore, intuitively, the EKF should outperform WLS if its underlying process model hypothesis is correct.

A comprehensive comparative analysis is presented in **Table 4**. The results from the table unmistakably indicate that the proposed method substantially improves the accuracy of DSSE when contrasted with conventional approaches. Particularly

Ref.	Proposed system	Method to solve	Model	No. of μ-PMU	Max error in bus voltage magnitude (%)	Max error in bus voltage angle (%)
[1]	Three-phase 13-bus distribution system	Modified Krawczyk operator algorithm	Mixed integer non-linear optimization model	4	2.510	3.281
[2]	Three-phase IEEE 33-bus distribution system	Safety barrier interior point optimization method	Non-linear optimization model	11	2.137	3.485
[3]	Three-phase IEEE 34-bus distribution system	Weighted least square method	Linear optimization model	12	2.568	—
[4]	Three-phase IEEE 37-bus distribution system	Hybrid PSO with chaotic gravitational search algorithm	Non-linear optimization model	13	1.508	1.602
[5]	Three-phase IEEE 69-bus distribution system	Safety barrier interior point optimization method	Non-linear convex optimization model	24	3.381	3.972
This ref.	Three-phase 85-bus distribution system	Weighted least square method	Nonlinear model with Taylor series approximations	30	1.980	2.750

Table 4.
Comparison of optimal solutions in different test systems.

noteworthy is the observation that the proposed method yields significantly lower maximum errors in bus voltage magnitude and voltage angles compared to alternative methodologies, emphasizing its superior performance.

6. Conclusion

Conclusion: This chapter presents a unique two-stage programming model. In the first stage, a strategy is formulated for the optimal placement of μ-PMUs, aiming to minimize installation costs while maximizing measurement redundancy and system observability, especially in the presence of PZINs. Following this, the second stage introduces an innovative DSSE model. This model employs the Taylor series approach for approximation, enabling the linearization of non-linear functions related to active and reactive power. The innovative nature of this approach ensures a comprehensive state estimation of the network, setting it apart from existing DSSE techniques. Following that, the linear DSSE problem is tackled using the WLS method to ascertain the states of the proposed system. Furthermore, the proposed method for distribution system state estimation, utilizing the EKF approach in addition to the WLS method, demonstrates high efficiency and robustness with minimal errors, as evidenced by the closely approximated estimated values to the actual values obtained from the backward-forward power flow method. Numerical analysis reveals the slight superiority of the EKF approach over WLS, attributed to its utilization of both past and present measurements, thus enhancing performance, particularly in scenarios where its underlying process model hypothesis holds. In essence, the suggested state estimation technique is put to the test in a practical case study conducted within the Matlab simulation environment. The results obtained from simulating an 85-bus distribution network highlight the approach's suitability for large-scale systems, underscoring its effectiveness and precision. Summarily, the conclusions derived from the simulation findings can be succinctly outlined as follows:

i. The proposed model is designed with the primary objective of minimizing the required μ-PMUs, leading to cost-effective installations, increased redundancy in measurements, and enhanced observability, particularly in the presence of PZINs and diverse contingencies.

ii. The results of the investigation demonstrate that the proposed model provides a dependable, resilient, sufficiently accurate, and stable state estimation in distribution systems incorporating DGs, all accomplished with a minimal quantity of μ-PMUs.

iii. The applicability of the proposed model extends seamlessly to any distribution system, regardless of its size, structure, load level, or the nature of distributed energy resources involved.

iv. This initiative is expected to elevate SA and ensure the dependable operation.

v. The strategic deployment of μ-PMUs has been demonstrated to enhance CPU timings, thereby optimizing the computational efficiency of the system.

vi. The results of the numerical experiments underscore the significant impact of the asymmetrical structure of distribution networks on real-world operations. This highlights the importance of taking into account techniques for single-phase state estimation.

Conflict of interest

The authors declare no conflict of interest.

Nomenclature

i,j	Bus index
n	Phase index
m	Measurement index
W, V	The price of both three and single-phase μ-PMU
X_j	The amount of binary variable is 1, if a three-phase μ-PMU exist at node j
P_j^n	The amount of binary variable is 1, if a single-phase μ-PMU exist at node j and phase n
a_{ij}^n	The (i,j) member of the incident matrix associated with phase n

Y_{ij}^n	An auxiliary binary variable
Z_j^n	The amount of binary parameter is 1, if a ZIN exists at node j and phase n
off_i^n	The amount of binary parameter is 1, if node i of phase n is off
ZB_j^n	The amount of binary parameter is 1, if a ZIN exist at the end node j and phase n
ZBR_j^n	Binary parameter, it is 1, for phase n of PZIN j that is also an end bus
R_i^n	Binary parameter, it is 1 for phase n of non-critical end bus i
w_u	Weight coefficient
e	Vector of measurement errors in m dimensions
z	Vector of measurements in m dimensions
x	Vector of states in n dimensions
h	Function for measurements
I_{ij}^m	Currents in different phases
ΔV_{ij}^m	Voltage drop between nodes i and j for phase m
Y	Matrix of admittances
S_{ij}^m	Flow of complex power
δ_i^m	Voltage angle at the ith bus and mth phase
V_i^m	Voltage magnitude at the ith bus and mth phase
$\overline{G}^m, \overline{B}^m$	Conductance and susceptance in the admittance matrix
P_{ij}^m	Active power flow in phase m
Q_{ij}^m	Reactive power flow in phase m
φ	Constraints of equality
ψ	Constraints of inequality
F_f	Objective function for the first stage
F_s	Objective function for the second stage

Author details

Arya Abdolahi[1]*, Selma Cheshmeh Khavar[2], Morteza Nazari-Heris[3] and Navid Taghizadegan Kalantari[1]

1 Azarbaijan Shahid Madani University, Tabriz, Iran

2 Amirkabir University of Technology, Tehran, Iran

3 Lawrence Technological University, Southfield, MI, USA

*Address all correspondence to: a.abdollahi@azaruniv.ac.ir

References

[1] Bindu S, Ushakumari S, Savier JS. Linear distribution system state estimation with integration of DG. Technology and Economics of Smart Grids and Sustainable Energy. 2021;**6**(1):1-7

[2] Sarada Devi MS, Yesuratnam G. Comparison of State Estimation Process on Transmission and Distribution Systems. In: Advances in Decision Sciences, Image Processing, Security and Computer Vision: International Conference on Emerging Trends in Engineering (ICETE). Vol. 2. Springer International Publishing; 2020. pp. 414-423

[3] Wang G, Giannakis GB, Chen J, Sun J. Distribution system state estimation: An overview of recent developments. Frontiers of Information Technology & Electronic Engineering. 2019;**20**(1): 4-17

[4] Nanchian S, Majumdar A, Pal BC. Three-phase state estimation using hybrid particle swarm optimization. IEEE Transactions on Smart Grid. 2017; **8**(3):1035-1045

[5] Ullah Z, Elkadeem MR, Wang S, Radosavljevic J. A novel PSOS-CGSA method for state estimation in unbalanced DG-integrated distribution systems. IEEE Access. 2020;**8**: 113219-113229

[6] Ye G, Nijhuis M, Cuk V, Cobben JFG. Incorporating network uncertainties in voltage dip state estimation. International Journal of Electrical Power & Energy Systems. 2019;**113**:888-896

[7] Pokhrel BR, Bak-Jensen B, Pillai JR. Integrated approach for network observability and state estimation in active distribution grid. Energies. 2019; **12**(12):2230

[8] Kong W, Yuang Y. Multi objective for PMU placement in compressed distribution network considering cost and accuracy of state estimation. Applied Sciences. 2019;**9**(7):1515

[9] Moshtagh S, Rahmani M. Robust hybrid state estimation for power systems utilizing phasor measurements units. Electric Power Systems Research. 2021;**196**:107195

[10] Shahriar M, Habiballah I, Hussein H. Optimization of phasor measurement unit (PMU) placement in supervisory control and data acquisition (SCADA)-based power system for better state-estimation performance. Energies. 2018; **11**(3):570

[11] Ni F, Nguyen PH, Cobben JFG, Van den Brom HE, Zhao D. Three-phase state estimation in the medium-voltage network with aggregated smart meter data. International Journal of Electrical Power & Energy Systems. 2018;**98**: 463-473

[12] Jin T, Shen X. A mixed WLS power system state estimation method integrating a wide-area measurement system and SCADA technology. Energies. 2018;**11**(2):408

[13] Zhang L, Lai K. A novel complex linear state estimator for smart power distribution systems: Methodology and implementation. International Journal of Electrical Power & Energy Systems. 2020;**123**:106312

[14] Tsitsimelis A, Antón-Haro C. A regularized state estimation scheme for a robust monitoring of the distribution grid. International Journal of Electrical Power & Energy Systems. 2020;**117**: 105621

[15] Müller HH, Castro CA, Dotta D. Allocation of PMU channels at substations for topology processing and state estimation. IET Generation, Transmission & Distribution. 2020; **14**(11):2034-2045

[16] de Oliveira BC, Pereira JLR, et al. Decentralized three-phase distribution system static state estimation based on phasor measurement units. Electric Power Systems Research. 2018;**160**: 327-336

[17] Alzate EB, Bueno-Lopez M, Xie J, Strunz K. Distribution system state estimation to support coordinated voltage-control strategies by using smart meters. IEEE Transactions on Power Systems. 2019;**34**(6):5198-5207

[18] Kabiri M, Amjady N, Shafie-khah M, Catalão JPS. Enhancing power system state estimation by incorporating equality constraints of voltage dependent loads and zero injections. International Journal of Electrical Power & Energy Systems. 2018;**99**:659-671

[19] Melo ID, Pereira JLR, Ribeiro PF, Variz AM, Oliveira BC. Harmonic state estimation for distribution systems based on optimization models considering daily load profiles. Electric Power Systems Research. 2019;**170**:303-316

[20] Yang H, Qiu RC, Chu L, Mi T, Shi X, Liu CM. Improving power system state estimation based on matrix-level cleaning. IEEE Transactions on Power Systems. 2020;**35**(5):3529-3540

[21] Adjerid H, Maouche AR. Multi-agent system-based decentralized state estimation method for active distribution networks. Computers & Electrical Engineering. 2020;**86**:106652

[22] Matsukawa Y, Watanabe M, Mitani Y, Othman ML. Multi-objective PMU placement optimization considering the placement cost including the current channel allocation and state estimation accuracy. Electrical Engineering in Japan. 2019; **207**(2):20-27

[23] Bandara WGC, Almeida D, Godaliyadda RI, Ekanayake MP, Ekanayake J. A complete state estimation algorithm for a three-phase four-wire low voltage distribution system with high penetration of solar PV. International Journal of Electrical Power & Energy Systems. 2021;**124**: 106332

[24] Niyaragh SMM, Irani AJ, Shayeghi H. Modeling of zero injection buses based to optimal placement of PMUs for full observability of power systems. Journal of Electrical Engineering & Technology. 2020;**15**(6): 2509-2518

[25] Ruben C, Dhulipala SC, Bretas AS, Guan Y, Bretas NG. Multi-objective MILP model for PMU allocation considering enhanced gross error detection: A weighted goal programming framework. Electric Power Systems Research. 2020;**182**:106235

[26] Trevizan RD, Ruben C, Rossoni A, Dhulipala SC, Bretas A, Bretas NG. μPMU-based temporal decoupling of parameter and measurement gross error processing in DSSE. Electricity. 2021; **2**(4):423-438

[27] Brandalik R, Wellssow W-H. Power system state estimation with extended power formulations. International Journal of Electrical Power & Energy Systems. 2020;**115**:105443

[28] Prasad S, Kumar DV. Optimal allocation of measurement devices for distribution state estimation using multiobjective hybrid PSO–krill herd

algorithm. IEEE Transactions on Instrumentation and Measurement. 2017;**66**(8):2022-2035

[29] Soltani Z, Ma S, Khorsand M, Vittal V. Simultaneous robust state estimation, topology error processing, and outage detection for unbalanced distribution systems. arXiv preprint arXiv:2105.10111. 2021

[30] Chandra A, Pradhan AK. Online voltage stability and load margin assessment using wide area measurements. International Journal of Electrical Power & Energy Systems. 2019;**108**:392-401

[31] Yildiz T, Acilan E, Gol M. A partitioning strategy for improved state estimation performance in ill-conditioned power systems with hybrid measurement set. Electric Power Systems Research. 2021;**201**:107515

[32] Kim B-H, Kim H. PMU optimal placement algorithm using topological observability analysis. Journal of Electrical Engineering & Technology. 2021;**16**(6):2909-2916

[33] Abdolahi A, Kalantari NT. State estimation of asymmetrical distribution networks by μ-PMU allocation: A novel stochastic two-stage programming. Electric Power Systems Research. 2022; **213**:108738

[34] Zhou Z, Zhang J, Liu P, Li Z, Georgiadis MC, Pistikopoulos EN. A two-stage stochastic programming model for the optimal design of distributed energy systems. Applied Energy. 2013;**103**:135-144

[35] Azarhooshang A, Sedighizadeh D, Sedighizadeh M. Two-stage stochastic operation considering day-ahead and real-time scheduling of microgrids with high renewable energy sources and electric vehicles based on multi-layer energy management system. Electric Power Systems Research. 2021;**201**: 107527

[36] Theodorakatos NP. Optimal phasor measurement unit placement for numerical observability using branch-and-bound and a binary-coded genetic algorithm. Electric Power Components and Systems. 2019;**47**(4–5):357-371

[37] Abdolahi A, Taghizadegan N, Banaei MR, Salehi J. A reliability-based optimal μ-PMU placement scheme for efficient observability enhancement of smart distribution grids under various contingencies. IET Science, Measurement & Technology. Oct 2021; **15**(8):663-680

[38] Helmi AM, Carli R, Dotoli M, Ramadan HS. Efficient and sustainable reconfiguration of distribution networks via metaheuristic optimization. IEEE Transactions on Automation Science and Engineering. 2021;**19**:82-98

[39] Salehi J, Abdolahi A. Optimal scheduling of active distribution networks with penetration of PHEV considering congestion and air pollution using DR program. Sustainable Cities and Society. 2019;**51**:101709

[40] Mohamed MA, Al-Sumaiti AS, Krid M, Awwad EM, Kavousi-Fard A. A reliability-oriented fuzzy stochastic framework in automated distribution grids to allocate μ-PMUs. IEEE Access. 2019;**7**:33393-33404

Chapter 5

Perspective Chapter: Approximate Kalman Filter Using M-Robust Estimate Dynamic Stochastic Approximation with Parallel Adaptation of Unknown Noise Statistics by Huber's M-Robust Parameter Estimator

Branko Kovačević, Zoran Banjac and Tomislav Unkašević

Abstract

The problem of designing a feasible adaptive M-robustified Kalman filter in a case of a thick-tailed Gaussian environment, characterized by impulsive noise-inducing observation and innovation outliers, and/or errors in mathematical model-inducing structural outliers, has been considered. Firstly, the time-varying criterion is used to generate a family of dynamic stochastic approximation algorithms. The M-robust estimate stochastic approximation is derived by minimizing the minimum variance criterion, the estimates of the latter being combined with the one-step minimum mean square error prediction to design M-robust estimate Kalman filter. Finally, the latter is combined with the Huber moving window M-robust parameter estimator of the unknown noises statistics, in parallel with the M-robust state estimation to design an adaptive M-robust estimate Kalman filter. Simulated maneuvering target tracking scenario revealed that the proposed adaptive M-robust estimate-based Kalman filter improves significantly the target estimation and tracking quality, being effective in suppressing multiple outliers with contamination degrees less than thirty percent. Moreover, the achieved improvement comes with additional computational efforts. However, these efforts are usually not significant enough to prevent real-time application.

Keywords: Kalman filter, adaptive filter, non-linear filter, non-Gaussian noise, impulsive noise, outliers, robust estimation, target tracking

1. Introduction

An optimal estimate of a stochastic variable, or a stochastic process, using noisy measurements, is the one that maximizes, or minimizes, a suitable performance index, the examples of the latter being least squares (LS), maximum a posteriori probability (MAP), maximum likelihood (ML), minimum variance (MV), minimum mean-square error (MMSE), least absolute value (LAV), etc. [1–5]. Moreover, the optimization criteria usually depend on suppositions upon the statistical performance of the system parameters and/or states that have to be estimated. One of the most important contributions to the estimation field, significant from both theoretical and practical perspectives, is the Kalman filter (KF), [1–5]. However, if the system dynamics, and adjoin measurements, are under the influence of severe nonlinear effects that may not be linearized adequately, and/or the noises are thick-tailed Gaussian distributed, KF deteriorates its performance, and even ceases to work [6, 7]. In such situations, it may be obtained bounded estimation error, in the statistical sense, by utilizing the dynamic stochastic approximation [8]. Therefore, the dynamic stochastic approximation performs quite well in various practical problems, involving optimization, state, and parameter estimation, as well as pattern recognition and classification, [9–12].

Additionally, in numerous practical situations, the real noises are thick-tailed Gaussian distributed, generating rare but high-intensity samples, named observation and innovation outliers, corrupting the mostly normally distributed measurements. The third type of outliers, named structural outliers, is caused by errors in a mathematical model, such as unmodeled system dynamics, time-varying bias, computational errors, etc. However, since KF is a linear signal processor that linearly transforms observations to calculate the system state updates, it is susceptible to outlying samples, being not robust. Therefore, it is of interest to design a robustified modification of the optimal KF, being able to manage a local not Gaussian noise vicinity, typified by impulsive disturbances inducing unavoidable outliers [13–18]. In such situations, robust methods of mathematical statistics produce appropriate tools to cope with outliers for reducing their influence on the estimation quality [19–26]. Moreover, the Huber's approximate maximum likelihood, or M-robust, approach is commonly applied by practitioners [21]. The history of research activities in the field of robust system parameter and/or state estimation through the application of different schemes, combining the Huber M-robust estimator and the LS, or KF, techniques, is pretty long, and many articles are devoted to these topics [17, 18, 27–37]. Thus, the emphasized solutions can be divided into two classes. The first one contains the batch processing, off-line, or non-recursive estimation procedures, in which the KF state estimation formulation is transformed to an adjoin parameter regression problem, the latter being resolved by utilizing the Huber M-robust approach [28–32]. The so attained robust procedure requires an iterated numerical method, such as the classical Newton method or its appropriate modification, as well as iterated reweighted LS [24, 25, 29–32]. Here, the robustness feature is obtained using simultaneous processing of predictions and observations, making the procedure efficient in reducing the effects of outliers, but at the cost of larger numerical efforts. Another class contains the recursive, sequential, or online parameter and state estimators, due to practical requirements for the fast computations in real-time signal processing. These estimators represent a good trade-off between the computational complexity and estimation quality [13–18, 33–37]. Particularly, it was proposed, in the recent past, an approximate KF, where the M-robust estimate dynamic stochastic approximation is used for the measurement update [17, 18]. Unfortunately, the M-robust scheme lacks sufficient adaptability, having a lower quality upon the pure normally distributed observations, but may also degrade its performance otherwise [17, 18, 27, 33–37]. Therefore, an adaptation to the local stochastic environment is needed. An appropriate approach to solve this problem, utilizing a combination of both the dynamic stochastic approximation and general time-varying M-robust criterion, together with the parallel Huber's moving window M-robust parameter estimator of the first and second-order unknown observation and state noises statistics, has been presented in this manuscript. The organization of the manuscript is the next. An overview of the KF and brief presentation of the robustness concepts are given in Chapter 2. Chapter 3 considers a design of the M-robust estimate KF. Chapter 4 is devoted to designing an adaptive M-robust estimate KF. The proposed approach combines the M-robust estimate KF and the Huber's moving window M-robust parameter estimator of unknown noise statistics. Simulation results that illustrate theoretical derivation, and demonstrate clearly the merits of emphasized adaptive M-robust estimate KF, using the real-life maneuvering target tracking example, are given in Chapter 5. Concluding remarks are presented in Chapter 6.

2. Overview of Kalman filtering technique

The optimal KF performances hold on to the recursive linear structure and the minimum variance procedure for calculating the gain sequence. Implementation of the optimal KF assumes both the system dynamics and adjoin measurement processes to be given by the linear, discrete-time, and state-space representation

$$x_{k+1} = F_k x_k + G_k w_k; y_k = H_k x_k + v_k \tag{1}$$

where x_k is the $n \times 1$ system state, y_k is the $r \times 1$ multidimensional observation, w_k is the $m \times 1$ state noise, and v_k is the $r \times 1$ observation noise, at the discrete-time, k. Furthermore, the sequences, $\{v_k\}$ and $\{w_k\}$, are the zero-mean white stochastic disturbances, being mutually uncorrelated, as well uncorrelated with the random initial state, x_0, with the mean value, m_0, and the covariance, P_0, that is

$$E\left\{\begin{bmatrix} w_k \\ v_k \end{bmatrix}\begin{bmatrix} w_k^T & v_k^T \end{bmatrix}\right\} = \begin{bmatrix} Q_k\delta_{kj} & 0 \\ 0 & R_k\delta_{kj} \end{bmatrix}; \; E\left\{\begin{bmatrix} w_k \\ v_k \end{bmatrix}[x_0 - m_0]^T\right\} = 0 \tag{2}$$

where $E\{\cdot\}$ is the mathematical expectation, and δ_{kj} is the Kronecker's delta symbol ($\delta_{kj} = 0$ for $k \neq j$ and $\delta_{kj} = 1$ for $k = j$). Here, F_k is the $n \times n$ state transition matrix, G_k is the $n \times m$ state noise transition matrix, and H_k is the $r \times n$ observation matrix.

Let $\hat{x}_{k/j} = E\{x_k|Y^j\}$, $(j = k-1, k)$, with $E\{\cdot|\cdot\}$ being the conditional mathematical expectation, represent the optimal LS estimate of the present state, x_k, upon the observation set $Y^j = \{y_i, i \leq j\}$, while $P_{k/j} = E\left\{\left(x_k - \hat{x}_{k|j}\right)\left(x_k - \hat{x}_{k|j}\right)^T\right\}$ designating the underlying error covariances. Then, the optimal KF relations are given by [1–5].

time update (*prediction*):

$$\hat{x}_{k+1/k} = F_k\hat{x}_{k/k}; P_{k+1/k} = F_k P_{k/k} F_k^T + G_k Q_k G_k^T \tag{3}$$

measurement update (*estimation, filtering*):

$$\hat{x}_{k/k} = \hat{x}_{k/k-1} + K_k\varepsilon_k; \; \varepsilon_k = y_k - H_k\hat{x}_{k/k-1} \tag{4}$$

$$K_k = P_{k/k-1}H_k^T S_k^{-1};\ S_k = H_k P_{k/k-1} H_k^T + R_k \tag{5}$$

$$P_{k/k} = (I - K_k H_k)P_{k/k-1}; K_k = P_{k/k} H_k^T R_k^{-1} \tag{6}$$

Thus, the filter can be initialized with

$$\hat{x}_{0/0} = E\{x_0\} = m_0 = 0;\ P_{0/0} = E\left\{[x_0 - m_0][x_0 - m_0]^T\right\} = P_0 \tag{7}$$

Here, the measurement residual, or innovation, ε_k in (4), is a zero-mean white stochastic sequence, with the variance S_k, in (5).

However, since the optimal KF linearly transforms observations, the state updates are not constrained, unless the system states are independent of the observations when the state updates are identically zero. In this sense, the linear optimal KF is susceptible to outliers, being not robust.

Thus, in the case of the Gaussian distributed observations contaminated by outliers, the two estimators are frequently used by practitioners [14, 29, 30]. In this sense, if the observation noise pdf is given by the δ-contaminated Gaussian pdf

$$CN(\cdot|\delta, \sigma_n^2, \sigma_0^2) = (1-\delta)N(\cdot|0, \sigma_n^2) + \delta N(\cdot|0, \sigma_0^2);\ 0 \le \delta < 1, \sigma_0^2 \gg \sigma_n^2 \tag{8}$$

then there are two possible procedures to handle outliers:

a. the standard KF treats the zero-mean Gaussian distributed observations with the increased known variance

$$R = (1-\delta)\sigma_n^2 + \delta\sigma_o^2 \tag{9}$$

b. the nominal KF ignores abnormality, or outliers, and simply treats the Gaussian distributed observations with the given nominal variance, $R = \sigma_n^2$.

In many cases of the non-Gaussian observations, the standard KF works fairly well, while the nominal one produces bad results, and even diverges [6, 7]. However, for complicated observation pdfs, and particularly if the non-Gaussian-ness is in a thick-tailed variety, as in (8), giving rise to observation and innovation outliers, the standard KF performance may also be quite poor [13–18, 28–33].

Formal mathematical definitions of robustness are given in the mathematical statistics [19–23], and the four such definitions exist. The two of them are ad hoc and data-oriented, named resistant and efficiency robustness, being preferable by the practitioners. Furthermore, an estimate is resistantly robust if it is insensitive both to a single outlying data point and the patchy, or grouped, ones. Additionally, some estimators are efficiently robust if they produce a high performance upon both the normally distributed measurements and normal ones under contamination, involving observation, innovation, and structural outliers [19–21]. Therefore, the practical robustness involves both the efficiency and resistant properties. The other two definitions, named min-max and qualitative robustness, have strong theoretical foundations [21, 23]. Thus, the qualitative robustness utilizes the Hampel's influence function, measuring a robust estimator's ability to cope with outliers [23]. On the other hand, the min-max robustness holds onto minimization of the estimator worst-case performance within the given distribution family to which the real observation noise distribution is confined [21]. Such optimization task is complex, and its exact solution may be derived only for a time-invariant model when the Fisher information is minimized [21, 27]. Thus, the robust min-max estimator reduces to the optimal ML one, whose likelihood function is defined by the worst-case pdf. On contrary, the Huber M-robust estimator is not quite the optimal ML one, but it approximates the last one to achieve both the efficiency and resistant robustness performances.

A simple and efficient concept to robustify the optimal KF can be developed using the Huber M-robust procedure. A version of this approach, based on both the one-step MMSE prediction and M-robust estimate dynamic stochastic approximation, is presented in the sequel.

3. M-robust estimate Kalman filtering

Although the dynamic stochastic approximation method is primarily designed to solve parameter regression problems, it may be extended to robust estimation of the dynamic system states, represented by (1), assuming the scalar measurements. The vectorial observations can be handled by considering components of the observation vector as the scalar observation, being used one at a time. Here is assumed that the observation vector components, in (1), are mutually uncorrelated scalar random variables, yielding a diagonal covariance, R in (2), A stable numerical algorithm to diagonalize a symmetric covariance matrix is the Choleskey factorization, or the UD decomposition [2]. In addition, due to simplicity, it uses the shortened notation $\hat{x}_{k|k-1} = \overline{x}_k$, $\hat{x}_{k|k} = \hat{x}_k$, $P_{k|k-1} = M_k$, and $P_{k|k} = P_k$. In this sense, one can define the time-varying functional of the scaled innovations, representing the M-robust optimization criterion

$$J_k(\overline{x}_k) = E\left\{\rho\left(\frac{\varepsilon_k(\overline{x}_k)}{s_k}\right)\middle|\overline{x}_k, Y^k\right\} \tag{10}$$

Here, $E\{\cdot|\overline{x}_k, Y^k\}$ is the conditional mathematical expectation given the current system state prediction, $\overline{x}_k$, and the measurement set, $Y^k = \{y_1, \dots, y_k\}$, where $\rho(\cdot)$ is Huber's robust score function

$$\rho(z) = \begin{cases} \Delta|z| - \Delta^2/2; & |z| \geq \Delta \\ z^2/2; & |z| < \Delta \end{cases}; \psi(\cdot) = \rho'(\cdot) = \min(|z|, \Delta)\,\mathrm{sgn}(z) \tag{11}$$

that has to cut off the outliers, while ψ is the influence function, that determines the estimator's ability to reduce the effects of different outlier types, including the observation, innovation, and structural outliers [17–19, 22–26]. Here, the saturation threshold, Δ, is a tuning parameter, providing for the required efficiency upon the pure Gaussian distributed observations, in (8) with $\delta = 0$.

Using step-by-step minimization of the M-robust criterion (10), one generates a family of the recursive gradient-type algorithms

$$\hat{x}_k = \overline{x}_k + \Gamma_k g_k(\overline{x}_k); g_k(\overline{x}_k) = -\nabla_{\overline{x}} J_k(\overline{x}_k) \tag{12}$$

where $\nabla_{\overline{x}} J_k(\cdot)$ is gradient vector of the optimization criterion in (10), while the innovation or measurement residual, ε, is given by

$$\varepsilon_k = \varepsilon(\overline{x}_k) = y_k - H_k\overline{x}_k; S_k = E\{\varepsilon_k\varepsilon_k^T\} = H_kM_kH_k^T + R_k; s_k = S_k^{1/2} \tag{13}$$

with S_k being the residual variance, with the corresponding standard deviation, s_k. Taking into account (10) and (11), one obtains the gradient vector expression

$$\nabla_{\overline{x}}J_k(\overline{x}_k) = \frac{\partial_k J_k(\overline{x}_k)}{\partial \overline{x}_k} = -\frac{1}{s_k}E\left\{\psi\left(\frac{\varepsilon_k}{s_k}\right)\bigg|\overline{x}_k, Y^k\right\}H_k^T \tag{14}$$

However, the expectation, in (14), is undeterminable and may be represented approximately by the current sample, producing the stochastic gradient relation

$$m_k = -\frac{1}{s_k}\psi\left(\frac{\varepsilon_k}{s_k}\right)H_k^T \tag{15}$$

Thus, by replacing (15) with (12), one gets

$$\hat{x}_k = \overline{x}_k - \Gamma_k m_k = \overline{x}_k + \frac{1}{s_k}\Gamma_k H_k^T\psi\left(\frac{\varepsilon_k}{s_k}\right) \tag{16}$$

with s_k being the normalizing, or scaling, factor, in (13).

The gain sequence, Γ_k in (16), has to supervise the convergence rate. Therefore, to link the M-robust dynamic stochastic approximation method (16), and the optimal KF, in (3)–(6), an additional approximate MV optimization criterion is adopted

$$J_1(\Gamma_k) = Trace\,P_k; P_k = E\{\tilde{x}_k(+)\tilde{x}_k^T(+)\}; \tilde{x}_k(+) = x_k - \hat{x}_k \tag{17}$$

with P_k being the underlying error covariance, while the matrix *Trace* is the sum of its diagonal components. The posed optimization problem, in (17), is a complex one that cannot be solved exactly, and a suboptimal solution has to be derived. Thus, the relation (16) can be rewritten in a recursive weighted LS form, being equal formally to the KF time-update (4), that is

$$\hat{x}_k = \overline{x}_k + K_k\varepsilon_k; K_k = s_k^{-2}\omega_k\Gamma_kH_k^T \tag{18}$$

with the robust weighting, normalizing, or penalty, term

$$\omega_k = \begin{cases} \dfrac{\psi(\varepsilon_k/s_k)}{\varepsilon_k/s_k} & \text{for } \varepsilon_k \neq 0 \text{ and } s_k \neq 0 \\ 1 & \text{for } \varepsilon_k = 0 \text{ and/or } s_k = 0 \end{cases} \tag{19}$$

The prediction error, $\tilde{x}_k(-)$, is given by (1) and (3), from which it follows

$$\tilde{x}_k(-) = x_k - \overline{x}_k = F_{k-1}\tilde{x}_{k-1}(+) + G_{k-1}w_{k-1} \tag{20}$$

where, due to (18), (13), and (1), the estimation error, $\tilde{x}_k(+)$, is determined by

$$\tilde{x}_k(+) = x_k - \hat{x}_k = \tilde{x}_k(-) - K_kH_k\tilde{x}_k(-) - K_kv_k \tag{21}$$

with $\{w_k\}$ and $\{v_k\}$ being the state and observation noises sequences of the zero-mean uncorrelated random variables, in (1) and (2). Under the initial guess (7), one concludes, by the mathematical induction, that the errors (20) and (21) are unbiased. Furthermore, under the assumption (2), the estimation errors and the noises are mutually uncorrelated, yielding

$$E\{\tilde{x}_k(-)\} = 0; E\{\tilde{x}_k(+)\} = 0; E\{\tilde{x}_k(+)w_k^T\} = 0; E\{\tilde{x}_k(-)v_k^T\} = 0 \tag{22}$$

Taking into account (20) and (22), one concludes that the covariance, M_k, is defined by the KF Eq. (3), yielding

$$M_k = E\{\tilde{x}_k(-)\tilde{x}_k^T(-)\} = F_{k-1}P_{k-1}F_{k-1}^T + G_{k-1}Q_{k-1}G_{k-1}^T \tag{23}$$

Moreover, it follows from (2), (21), and (22) that the error covariance, P_k in (17), is given by

$$P_k = E\{\tilde{x}_k(+)\tilde{x}_k^T(+)\} = M_k - K_kH_kM_k - M_kH_k^TK_k^T + R_kK_kK_k^T \tag{24}$$

By substituting (18) into (24), the relation (17) reduces to

$$J_1(\Gamma_k) = Trace\, M_k - 2s_k^{-2}\omega_k Trace\,\Gamma_k H_k^T H_k M_k + s_k^{-4}\omega_k^2 R_k Trace\,\Gamma_k H_k^T H_k \Gamma_k \tag{25}$$

After applying the partial derivation, based on the rules to partial derivation of trace of matrices product, [1, 4].
$\frac{\partial}{\partial A} TraceBAC = B^TC^T$; $\frac{\partial}{\partial A} TraceABA^T = 2AB$ if $B = B^T$
one obtains from (25)

$$\frac{\partial J_1(\Gamma_k)}{\partial \Gamma_k} = -2s_k^{-2}\omega_k M_k H_k^T H_k + 2s_k^{-4}\omega_k^2 R_k \Gamma_k H_k^T H_k = 0 \tag{26}$$

Additionally, due to (11), the term ω_k, in (19), can be approximated by only the zero and unity values, yielding $\omega_k{}^2 \approx \omega_k$, from which it follows

$$\frac{\partial J_1(\Gamma_k)}{\partial \Gamma_k} = -2s_k^{-2}\omega_k\left(M_k - s_k^{-2}R_k\Gamma_k\right)H_k^T H_k = 0 \tag{27}$$

Moreover, from (27) and (13), one gets

$$\Gamma_k = s_k^2 R_k^{-1} M_k = \left(R_k^{-1}H_kM_kH_k^T + 1\right)M_k \approx M_k;\;\; s_k^2R_k^{-1}\approx 1 \tag{28}$$

The approximation of Γ_k by M_k is based on the following reasoning: in saturation region of the ψ nonlinearity (11), indicating the outliers presence, the observation noise variance, R_k in (9), has a huge value. On contrary, in linear segment of the ψ nonlinearity, indicating the outliers' absence, the M-robust filter (18) reduces to the optimal KF, (3)–(6), yielding a low prediction error covariance M_k. Applying (28), (18), and (24), one gets

$$P_k = M_k - 2s_k^{-2}\omega_k M_k H_k^T H_k M_k + R_k s_k^{-4}\omega_k^2 M_k H_k^T H_k M_k \approx M_k - s_k^{-2}\omega_k M_k H_k^T H_k M_k \tag{29}$$

Taking into account (18), the expression (29) reduces to the optimal KF Eq. (6), that is

$$P_k = (I - K_kH_k)M_k; \Gamma_k = M_k; K_k = s_k^{-2}\omega_k M_k H_k^T; K_k = \omega_k P_k H_k^T R_k^{-1} \tag{30}$$

Additionally, derivation of the fourth equation for the gain K_k, in (30), is based on the matrix inversion lemma [1, 4]. Thus, by substituting S_k from (13) into the third equation in (30), one obtains

$$K_k = \omega_k M_k H_k^T \left(H_k M_k H_k^T + R_k\right)^{-1} \tag{31}$$

Furthermore, by replacing (31) with the first equation in (30), and taking into account the approximation $\omega_k^2 \approx \omega_k$, one obtains

$$\omega_k P_k = \omega_k M_k - \frac{\omega_k M_k H_k^T H_k \omega_k M_k}{H_k \omega_k M_k H_k^T + \omega_k R_k} \tag{32}$$

After application of the matrix inversion lemma, stating that the relation.

$$P_2 = P_1 - P_1 H^T \left(H P_1 H^T + R\right)^{-1} H P_1$$

can be represented in an alternative form

$$P_2^{-1} = P_1^{-1} + H^T R^{-1} H$$

the relation (32) reduces to

$$P_k^{-1} = M_k^{-1} + H_k^T R_k^{-1} H_k; P_k \left(M_k^{-1} + H_k^T R_k^{-1} H_k\right) = I \tag{33}$$

where I is an identity matrix. Using (33), the relation (31) can be rewritten as

$$K_k = \frac{P_k \left(M_k^{-1} + H_k^T R_k^{-1} H_k\right) \omega_k M_k H_k^T}{H_k M_k H_k^T + R_k} = \omega_k P_k H_k^T R_k^{-1} \tag{34}$$

being identical to the fourth relation, in (30). In accordance with the optimal KF, the derived M-robust estimate KF has the next recursive form.

time update (prediction), (3), (23):

$$\overline{x}_{k+1} = F_k \hat{x}_k + G_k \overline{w}_k; M_{k+1} = F_k P_k F_k^T + G_k Q_k G_k^T; \overline{w}_k = E\{w_k\} \tag{35}$$

measurement update (estimation, filtering), (11), (13), (18), (19), (30):

$$\hat{x}_k = \overline{x}_k + K_k \varepsilon_k; \;\; \varepsilon_k = y_k - H_k \hat{x}_k - \overline{v}_k; \overline{v}_k = E\{v_k\} \tag{36}$$

$$S_k = s_k^2 = E\left\{\varepsilon_k \varepsilon_k^T\right\} = H_k M_k H_k^T + R_k; \omega_k = \begin{cases} \dfrac{\psi(\varepsilon_k / s_k)}{\varepsilon_k / s_k} & \text{for } \varepsilon_k \neq 0 \text{ and } s_k \neq 0 \\ 1 & \text{for } \varepsilon_k = 0 \text{ and/or } s_k = 0 \end{cases} \tag{37}$$

$$K_k = s_k^{-2} \omega_k M_k H_k^T; P_k = (I - K_k H_k) M_k; K_k = \omega_k P_k H_k^T R_k^{-1} \tag{38}$$

where, due to the assumption (2), $\overline{w}_k = \overline{v}_k = 0$.

There is the same ad hoc logic in the M-robustified KF gain calculation, in (34) or (38), and calculation of the KF gain, in (6). Thus, taking into account that the weighted factor ω, in (19) and (37), approximates the slope, or the first derivative, of the ψ nonlinearity (11), one concludes that $\omega(z) \approx 1$ if $|z| \leq \Delta$, while $\omega(z) \approx 0$ if $|z| > \Delta$. In this sense, for those measurements being confined to the linear region of ψ non-linearity, indicating the Gaussian distributed observations, value of the factor ω is closed to the unity value, and the expression (34) approximates properly the optimal KF gain, in (6). Furthermore, upon the saturation region of ψ non-linearity, indicating the outliers' existence, the factor ω is zero-valued, resulting in a low robust gain K, in (34), the value of the latter being produced a low value of the state update, in (18) and (28). This, in turn, reduces the effects of multiple outliers.

The application of the KF needs the state and observation noise statistics, in (1) and (2), to be known in advance. However, these requirements may be not fulfilled in practice, due to modeling errors, inducing structural outliers, as well the presence of innovation and observation outliers, [17, 18, 28–30]. A suitable approach to designing an adaptive modification of the emphasized M-robust estimate KF, in (35)–(38), based on the Huber moving window M-robust parameter estimator of unknown disturbances statistics, is considered in the next chapter.

4. Adaptive M-Robustifying Kalman filtering with unknown noise statistics

In various practical examples, components of the vector-valued measurements may be processed one at a time, so that the observations, in (1), become scalar random variables. Since the true system state is unknown, the observation noise sample, v_k, cannot be estimated by (1). However, an intuitive approximation of the observation noise sample, v_k, is given by the scalar random variable

$$r_k = y_k - H_k\hat{x}_k \tag{39}$$

where $\hat{x}_k$ is the M-robust state estimate, in (35)–(38). Taking into account (1), (2), and (22), one gets further

$$r_k = H_k(x_k - \hat{x}_k) + v_k = H_k\tilde{x}_k + v_k; \bar{r}_k = E\{r_k\} = E\{v_k\} = \bar{v}_k \tag{40}$$

where $\tilde{x}_k$ is the robust estimation error, in (17). In addition, taking into account (2), (22), and (40), the variance of the real scalar random variable, r_k, is given by

$$V_r(k) = E\left\{[r_k - \bar{r}_k][r_k - \bar{r}_k]^T\right\} = H_k E\{\tilde{x}_k\tilde{x}_k^T\}H_k^T + E\left\{[v_k - \bar{r}_k]^2\right\} = H_k P_k H_k^T + R_k \tag{41}$$

where P_k is the error covariance matrix, in (17). Therefore, if the first and second-order moments, $\bar{r}_k$ and $V_r(k)$, are somehow estimated, say by $\hat{r}_k$ and $\hat{V}_r(k)$, one can estimate further $\bar{v}_k$ and R_k, from (40) and (41), respectively. In this sense, one obtains the following estimates for the observation noise statistics (men value and variance)

$$\hat{v}_k = \hat{r}_k; \hat{R}_k = \left|\hat{V}_r(k) - H_k P_k H_k^T\right| \tag{42}$$

where P_k is generated by the M-robustified KF, (35)–(38). Here, $|\cdot|$ denotes the absolute value to prevent the negative variance estimates, due to possible numerical errors in applications.

Let us assume further, due to simplicity and clarity, that the state noise sample, w_k in (1), is a scalar random variable, as it is frequently fulfilled in tracking applications, [38–40]. In addition, if the system under consideration is a single input - single output (SISO), and the corresponding state-space model (1) is presented in the controllable canonical form, G in (1) is the $n \times 1$ column vector, with zero-valued components, except the last one that is unity valued, [4]. Thus, the random input, or state noise sample, w_k, is a real scalar random variable. Moreover, the state noise sample, w_k, can be estimated from the state Eq. (1), when the unknown random states, x_{k+1} and x_k, are replaced by the two successive robust estimates, $\hat{x}_{k+1}$ and $\hat{x}_k$, in (35)–(38). A such obtained state noise sample estimate represents a scalar random variable

$$q_k = T_k(\hat{x}_{k+1} - F_k\hat{x}_k); T_k = \left(G_k^T G_k\right)^{-1} G_k^T; G_k^T G_k = 1 \tag{43}$$

The relation (43) can be rewritten in an alternative form

$$q_k = T_k(\hat{x}_{k+1} - F_k\hat{x}_k - x_{k+1} + x_{k+1}) \tag{44}$$

which, after replacing (1) with (44) reduces to

$$q_k = T_k(-\tilde{x}_{k+1} - F_k\tilde{x}_k + G_k w_k); \tilde{x}_k = x_k - \hat{x}_k \tag{45}$$

Taking into account (1), (2), and (22), further follows that the unknown mean value, $\overline{w}_k$, of the state noise, w_k, is equal to the mean value, $\overline{q}_k$, of the random variable, q_k, that is

$$\overline{q}_k = E\{q_k\} = T_k G_k E\{w_k\} = \overline{w}_k; T_k G_k = 1 \tag{46}$$

Moreover, variance of the random variable, q_k in (43), is defined by

$$V_q(k) = E\left\{[q_k - \overline{q}_k]^2\right\} = E\left\{[q_k - \overline{q}_k][q_k - \overline{q}_k]^T\right\} \tag{47}$$

Bearing in mind (1), (2), and (22), and by substituting (45) and (46) into (47), one gets

$$\begin{aligned} V_q(k) = T_k\big[&P_{k+1} - F_k E\{\tilde{x}_k\tilde{x}_{k+1}^T\} - G_k E\{[w_k - \overline{w}_k]\tilde{x}_{k+1}^T\} - E\{\tilde{x}_{k+1}\tilde{x}_k^T\}F_k^T \\ &+ F_k P_k F_k^T - E\{\tilde{x}_{k+1}[w_k - \overline{w}_k]\}G_k^T + G_k E\left\{[w_k - \overline{w}_k]^2\right\}G_k^T\big]T_k^T \end{aligned} \tag{48}$$

Furthermore, by substituting the prediction error, in (20), for the estimation error, in (21), one obtains the estimation error recursion

$$\tilde{x}_{k+1} = (I - K_{k+1}H_{k+1})F_k\tilde{x}_k + (I - K_{k+1}H_{k+1})G_k[w_k - \overline{w}_k] - K_{k+1}[v_{k+1} - \overline{v}_{k+1}] \tag{49}$$

from which it follows

$$F_k E\{\tilde{x}_k\tilde{x}_{k+1}^T\} = F_k P_k F_k^T (I - K_{k+1}H_{k+1})^T; E\{\tilde{x}_{k+1}[w_k - \overline{w}_k]\} = (I - K_{k+1}H_{k+1})G_k Q_k \tag{50}$$

Furthermore, by substituting (50) into (48), together with using the expressions for the error covariances, M_k in (23), and, P_k in (30), one gets

$$V_q(k) = T_k\left(-P_{k+1} + F_k P_k F_k^T\right)T_k^T + Q_k \tag{51}$$

Therefore, if $\hat{q}_k$ and $\hat{V}_q(k)$ denote some estimates of the unknown mean value, $\overline{q}_k$ in (46), and the variance, $V_q(k)$ in (51), respectively, one can estimate the unknown first and second-order state noise statistics, $\overline{w}_k$ and Q_k, by the following equations

$$\hat{w}_k = \hat{q}_k; \hat{Q}_k = \left|\hat{V}_k(k) - T_k\left(-P_{k+1} + F_k P_k F_k^T\right)T_k^T\right| \tag{52}$$

Relations (42) and (52) are still valid for the KF, in (3)–(6), since the expressions for the state prediction, $\overline{x}_k$, prediction error covariance, M_k, state estimation update, $\hat{x}_k$, and estimation error covariance, P_k, are given by the identical equations to the M-robust estimate KF, in (35)–(38). Thus, the derived relations, (42) and (52), can be applied to designing not only an adaptive M-robustified KF but also to designing an adaptive linear optimal KF.

Next, to calculate the unknown observation noise statistics, in (42), as well as the unknown state noise statistics, in (52), a scalar constant parameter estimation problem has to be considered. Thus, let us consider a real scalar random variable, $\mathfrak{R}$, generating the independent and identically distributed (iid) samples, $r_i, i = k - L + 1, \dots, k$, in (39), being stored in a one-step moving window of the size L at the present stage, k. Using the empirical samples, the mean value, or the location parameter, $\overline{r}_k = m_{\mathfrak{R}}$, and the variance, $V_r(k) = V_{\mathfrak{R}}$, related to the random variable, $\mathfrak{R}$, has to be calculated at each stage, k. Unfortunately, analogously to the linear optimal KF, the commonly used sample mean and sample variance are not robust, [19–22]. Therefore, the Huber moving window M-robust parameter estimator is proposed as a robust alternative to the sample-mean and variance estimates. The Huber M-robust estimate, $\hat{r}_k$, of the unknown mean value, $\overline{r}_k$, minimizes the robust index of goodness, being given by the experimental sum

$$J_{ME}(\hat{r}_k) = \frac{1}{L}\sum_{i=k-L+1}^{k} \rho\left(\frac{r_i - \hat{r}_k}{\hat{d}_k}\right) \tag{53}$$

where $\{r_i\}$ is the given sequence of observations, in (39), covering the last L samples related to the current stage, k, and $\rho(\cdot)$ is the Huber's loss function, in (11), [21]. Thus, the M-robust estimate, $\hat{r}_k$ in (53), represents a solution of the nonlinear algebraic equation, obtained from the optimality condition

$$\left.\frac{dJ_{ME}(r)}{dr}\right|_{r=\hat{r}_k} = \sum_{i=k-L+1}^{k} \psi\left(\frac{r_i - \hat{r}_k}{\hat{d}_k}\right) = 0; \psi(x) = \frac{d\rho(x)}{dx} \tag{54}$$

where ψ is the saturation nonlinearity, in (11). The scaling factor, $\hat{d}_k$, provides for the scale-invariant characteristic of the M-robust location parameter estimator, in (53) and (54). Although ad hoc, a popular robust estimate, $\hat{d}_k$, is the median of the absolute median deviation (MAD)

$$\hat{d}_k = median\{r_i - median(r_i)\}/0.6745; i = k - l + 1, \dots, k \tag{55}$$

being defined on a one-step moving window of the size l [19, 21–25]. The divisor 0.6745 is taken since upon large enough sample size, l, and if the samples belong to the

Gaussian pdf, then $\hat{d}_k$ is equal approximately to the true variance, $\sigma^2_{\Re} = V_{\Re}$, [21]. Furthermore, the proposed procedure of choosing $\hat{d}_k$ indicates a suitable choice of the saturation threshold, Δ, in (11). Since, $\hat{d}_k \approx \sigma_{\Re}$, the free parameter, Δ, is commonly chosen to be some number about 1.5. Such estimate is named the Huber 1.5-M-robust parameter estimator [21]. The Eq. (54) is non-linear, and an iterated numerical algorithm is requested [21, 24, 25]. However, an approximate weighted LS approach results in a one-step suboptimal solution

$$\sum_{i=k-L+1}^{k} \omega_{i0}(r_i - \hat{r}_k) \approx 0; \hat{r}_k = \hat{r}_L = \frac{\sum\limits_{i=k-L+1}^{k} \omega_{i0} r_i}{\sum\limits_{i=k-L+1}^{k} r_i}; \omega_{i0} = \begin{cases} \dfrac{\psi\left(\dfrac{r_i - \hat{r}_0}{\hat{d}_k}\right)}{\dfrac{r_i - \hat{r}_0}{\hat{d}_k}}; & \hat{d}_k \neq 0 \text{ and } r_i \neq \hat{r}_0 \\ 1; & \hat{d}_k = 0 \text{ and/or } r_i = \hat{r}_0 \end{cases} \tag{56}$$

with $\hat{r}_0$ being an arbitrary initial guess [25, 29, 30]. Particularly, when $\hat{d}_k = \sqrt{V_{\Re}}$, the asymptotic variance of M-robust location parameter estimator (53), (54) is given by [21].

$$V_a = \lim_{L\to\infty} E\left\{L(\hat{r}_L - \bar{r})^2\right\} = \frac{V_{\Re} E\left\{\psi^2\left[\frac{r_i - \bar{r}}{\sqrt{V_{\Re}}}\right]\right\}}{E^2\left\{\psi'\left[\frac{r_i - \bar{r}}{\sqrt{V_{\Re}}}\right]\right\}}; \bar{r} = m_{\Re}; V_r = V_{\Re} \tag{57}$$

A natural estimate of V_a, in (57), is obtained when the undeterminable expectation is approximated by the average sum, together with the MAD estimator (55) for the scaling factor calculation at each stage, k. Thus, the unknown variance, $V_r(k)$ in (42), may be estimated by the Huber's moving window M-robust parameter estimator

$$\hat{V}_r(k) = \frac{\hat{d}_k^2 \left\{\frac{1}{L}\sum_{i=k-L+1}^{k} \psi^2\left[\frac{r_i - \hat{r}_k}{\hat{d}_k}\right]\right\}}{\left\{\frac{1}{L}\sum_{i=k-L+1}^{L} \psi'\left[\frac{r_i - \hat{r}_k}{\hat{d}_k}\right]\right\}^2} \tag{58}$$

Moreover, if $\psi(\cdot)$ is the linear function, the variance estimate (58) reduces to the sample variance.

The M-robust parameter estimator, in (55), (56), and (58), is used to calculate the nuisance estimates, $\hat{r}_k$ and $\hat{V}_r(k)$, the estimates of the latter are used to calculate the requested observation noise mean value, $\hat{v}_k$, and the variance, $\hat{R}_k$, from (42). The state noise sample estimates, q_k in (43), has to be managed simultaneously with the observation noise samples, r_k in (39). Thus, an additional one-step sliding frame of the size L has to be used, covering the last L samples, q_i, $i = k - L + 1, \dots, k$, at each stage, k. Again, the Huber moving window M-robust parameter estimator, in (55), (56), and (58), is used to estimate the nuisance parameters, $\hat{q}_k$ and $\hat{V}_q(k)$, the estimates of the latter are applied to calculating the required state disturbances statistics, $\hat{w}_k$ and $\hat{Q}_k$, from (52). The proposed adaptive estimation scheme needs an additional memory and shifting operations upon the sample estimates, r_i and q_i, within the given sliding data frames of the length L. Thus, the samples r_{k-L} and q_{k-L} are dropped when the current samples, r_k and q_k, are stored in the current sliding data frame.

It should be noted that the robust MAD estimator, in (55), may be also used to estimate the variances, $\hat{V}_r(k)$ and $\hat{V}_q(k)$, but the robust estimator (58) represents a better solution at the expense of increased computational requirements. Another alternative estimator to calculate robustly the variance may be found in the statistical literature [19–26]. These solutions use primarily some influence function, being non-decreasing near the origin, but decreasing toward zero from the origin, named redescending influence function. However, the underlying M-robust estimators may not have a unique solution, requiring a brief choice of the initial conditions, being generated, for example, by the Huber's robust estimator. Furthermore, during the filter initial steps, the disturbance sample estimates, (39) and (43), may represent bad indicators of the stochastic surroundings. Therefore, a fading memory method, weighting the successive disturbance sample estimates, r_k or q_k, by the growing factor

$$g_k = (k - L)(k - L - 1) \dots (k - L - \gamma)/k^{\gamma}; k = L, L + 1, \dots; \lim_{k \to \infty} g_k = 1 \tag{59}$$

can be applied to delay the individual noise sample estimates, r_k and q_k, over the first γ steps [15, 30]. Moreover, during the filter initialization, the first L noise samples, r_k and q_k, may be generated by using the Gaussian random number generator, with zero-mean and the corresponding variances, $\hat{R}_1$ and $\hat{Q}_0$, the values of the latter being used to define the fixed scaling factors at the initial data frames. It should be noted that in most applications, the mean values of observation and state noises are zero-valued, and there is no need to estimate these noise statistics. In such a case, the mean value estimates $\hat{r}_k$ and $\hat{q}_k$, in (55) and (56), are used only to calculate the variance estimates, $\hat{V}_k$ and $\hat{V}_q$, by (58), the estimates of the latter are used to calculate the required observation and state noise variances, using (42) and (52), respectively.

A theoretical convergence analysis of the emphasized M-robustified KF algorithm, either adaptive or not, is complex in the technical sense, owing to both a nonlinear scaled innovation processing and a time-varying multivariable dynamic system model. Moreover, drawing appropriate conclusions to sensitivity on multiple outliers, involving the observation, innovation, and structural ones, as well as choice of the initial conditions and the form of nonlinear influence function, is exceedingly challenging. Therefore, comprehensive Monte Carlo simulations are needed to refine the characteristics of the proposed adaptive M-robust estimate based KF and to bring appropriate conclusions.

5. Simulation results

A requirement for simple filters in real-time applications implies the desirability of uncoupled filtering, where independent tracking is performed in each coordinate (x, y, z) of the Cartesian Coordinates System (CCS). Particularly, three independent KF in the CCS coordinates are utilized, each of them being based on the three-dimensional state (position, velocity, and acceleration) representation, in (1). The first step in simulations is to generate the nominal target trajectory. The kinematic equations of motion, with respect to an inertial reference system, produce the true state variables, in (1), when a desirable acceleration profile, being a piece-wise constant over the sampling intervals, is adopted [40]. The so-obtained target trajectory is used in the presented simulations, and **Figure 1** depicts the actual target trajectory projections in each of the CCS planes xOy, yOz and xOz, respectively.

Because of a similar behavior of the actual components of three-dimensional state vector in each of the CCS coordinates, only the z-coordinate is represented in **Figure 2**, where z_k, $\dot{z}_k$, and $\ddot{z}_k$ are the position, velocity, and acceleration of a target, at the k-th scan. Proceeding from a thick-tailed Gaussian behavior of the position measurement probability distribution function, associated with the target glint noise spikes, inducing innovation and observation outliers, the observation noise sequence, $\{v_k\}$, is generated as the samples from the δ-contaminated normal pdf, in (8). The random noise sample, v_k, is produced by taking firstly a uniformly distributed sample, u, in the [0,1] interval. Thus, $u \leq \delta$, the sample, v_k, is produced by the contaminating normal pdf, $N(\cdot|0, \sigma_0^2)$, in other ways, the sample, v_k, belongs to the nominal normal pdf, $N(\cdot|0, \sigma_n^2)$. A typical heavy-tailed glint noise record is given in **Figure 3**.

The position measurements, in (1), are generated at each scan, k, by adding the previously generated observation noise sample, v_k, to the true target projection on the CCS coordinate of interest. For example, the noise sample, v_k, has to be added to the true z_k position, in **Figure 2**, assuming that the z-CCS coordinate is concerned.

The filter-world equations of motion include only the linear discrete-time state-space representation with only the position measurement, in (1). Thus, the corresponding system matrices take the form

$$F = F(T) = \begin{bmatrix} 1 & T & T^2/2 \\ 0 & 1 & T \\ 0 & 0 & 1 \end{bmatrix}; G = \begin{bmatrix} 0 \\ 0 \\ 1 \end{bmatrix}; H = \begin{bmatrix} 0 & 0 & 0 \end{bmatrix} \tag{60}$$

Here, T represents the constant period of sampling, and $T = 4s$ is used.

Radar observations commonly involve the range, r, elevation, β, and azimuth, α, angles, because the measurement errors are not coupled in the Spherical Coordinate System (SCS). Moreover, these measurements, when transformed to CCS, are no longer uncoupled, so the use of the three independent and uncoupled three-dimensional state KF, based on the model (1) and (60), in each of the CCS coordinates, may lead to less accurate filtering than it would occur if the coupled nine-

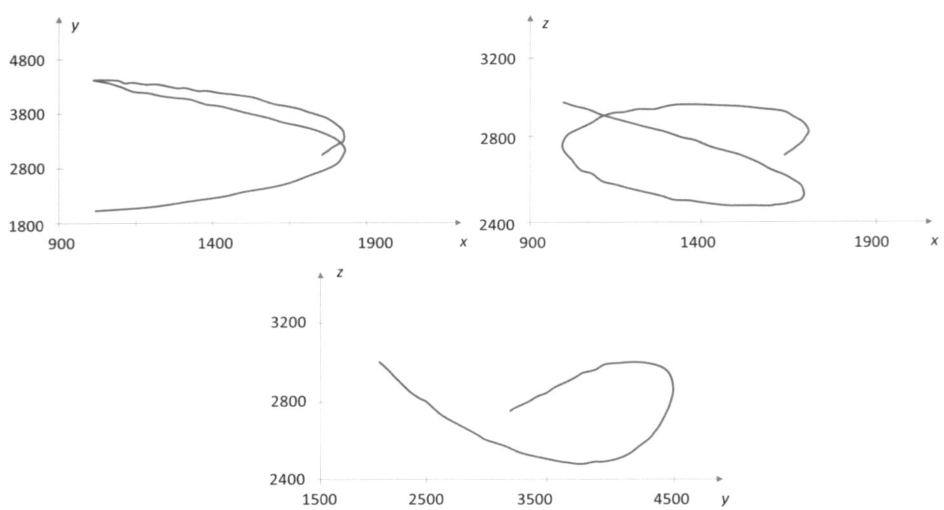

Figure 1.
The actual target trajectory in the CCS planes (x, o, y), (x, o, z), (y, o, z).

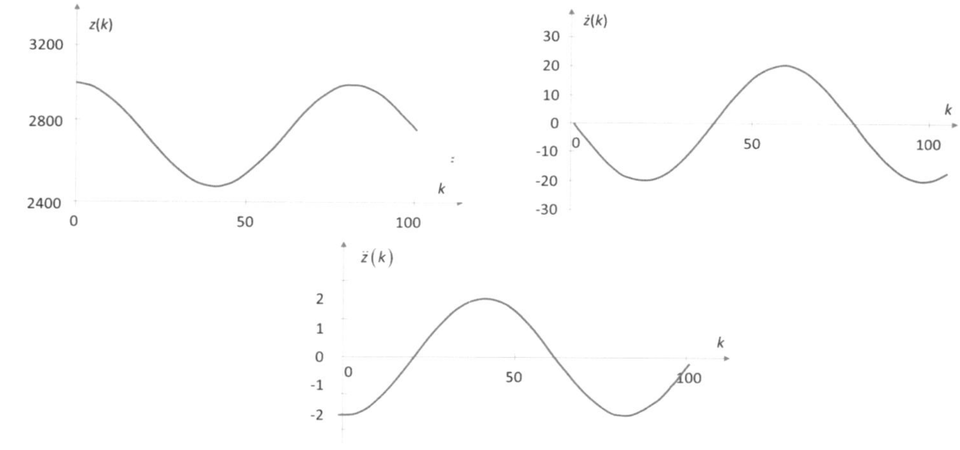

Figure 2.
The actual components of the three-dimensional state (position, velocity, and acceleration) in the z-CCS coordinate.

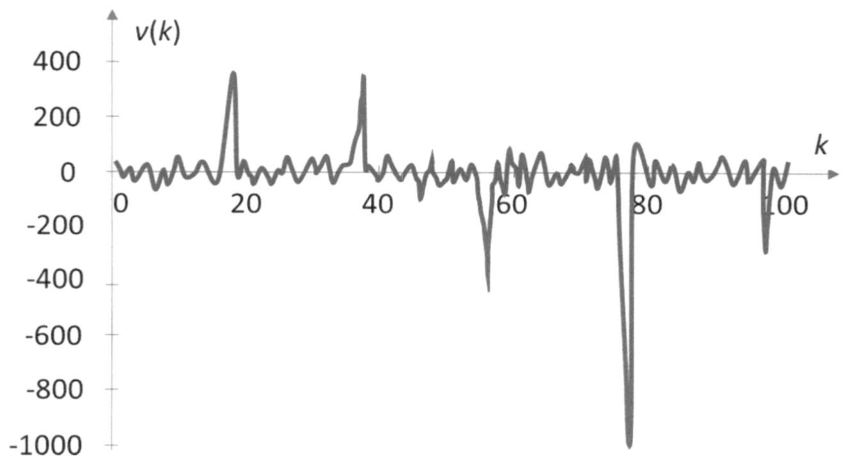

Figure 3.
Typical thick-tailed Gaussian glint noise record ($\delta = 0.05, \sigma_n = 1, \sigma_o = 90$).

dimensional state KF is used. Thus, the emphasized procedure is a good balance between tracking ability and computational requirements, providing also a meaningful test to tracking quality in the impulsive noise vicinity inducing innovation and observation outliers, as well significant errors of dynamic system model, due to a target maneuver, inducing structural outliers. The state-space model, (1) and (60), is named the constant acceleration (CA) one [40].

The implementation of the KF, robustified or not, requires the filter initialization, and a priori knowledge of the state and observation noises statistics. The optimal KF initialization, in (7), is based on the optimal estimates of zero-mean states, in (1), when the observations are not available. Thus, the adopted initial guesses, $m_0 = 0$ and P_0 in (7), are the steady-state values. However, such initialization approach cannot be used for a tracking problem, because the state vector does not reach a steady state covariance matrix, P_0. Namely, the cumulative effects of many acceleration perturbations affect the state vector component variances to grow with the time index, k, about their nominal values [39, 40]. Thus, a suboptimal and reasonable ad-hoc initialization can be developed using the two available position sensor measurements, in (1), yielding

$$\hat{x}_{0|0} = \hat{x}_0 = \begin{bmatrix} y_2 \\ \frac{y_2 - y_1}{T} \\ 0 \end{bmatrix}; \; P_{0|0} = P_0 = diag\{\sigma_1^2, \sigma_2^2, \sigma_3^2\} \tag{61}$$

Here, uncertainty in the initial three-dimensional state vector, in (1), is reflected in initial values of the covariance matrix, P_0 in (61), where σ_1, σ_2, and σ_3 denote the standard deviations of position, velocity, and acceleration in the three-dimensional state space model, (1) and (60). A reasonable choice is $\sigma_1 = 100m$, $\sigma_2 = 50m/s$, and $\sigma_3 = 10m/s^2$, [40].

The first-order statistics of state and observation noises are commonly zero-valued that is, $\overline{v} = \overline{w} = 0$ in (35) and (36). The state noise variance, Q, represents a random motion, maneuver, or acceleration entering a system between the sampling intervals. Starting from the state space model, in (1) and (60), the variance Q can be calculated by adopting the first-order Gauss-Markov state noise process, with the exponential correlation coefficient, assuming that $T \ll \tau_m$ where τ_m is the target maneuver time constant. Applying further the Singer's approach, the random target acceleration can be generated from the mixed pdf, representing a combination of the uniform pdf, $f(u)$, on the interval $[-A_m \; A_m]$, with A_m being the maximal target acceleration in both directions, and the three Dirac impulses $P_m\delta(u \pm A_m)$, $P_0\delta(u)$. Here, P_m is the probability of maximal accelerations, $\pm A_m$, while P_0 is the probability of zero-valued acceleration. Starting from such mixed pdf, one can calculate the random maneuver variance, σ_m^2, from which it follows the state noise variance [40].

$$Q = E\{w_k^2\} = 2T\sigma_m^2/\tau_m; \sigma_m^2 = A_m^2(1 + 4P_m - P_0)/3 \tag{62}$$

Particularly, by assuming $P_m = 0.1; P_0 = 0.3; A_m = 2g/3; T = 1s; \tau_m = 10s$ one gets $\sigma_m \approx g$, and $Q \approx = 0.45g$, where the gravity constant $g = 9.81m/s^2$.

For the limiting case, $T \ll \tau_m$, a suitable simplification to the two-dimensional-state KF becomes valid, since the KF gains adjoin to the acceleration estimates take low values, [29, 40]. However, the state noise must be taken to replace the omitted acceleration term, and such model is called the constant velocity (CV) one. Additionally, the limiting case, $T \ll \tau_m$, is not the nominal one, and usually $T \geq \tau_m$, [40]. Thus, the variance, Q, may be chosen in an ad-hoc manner, as an average Monte-Carlo value on the tracking interval of interest, [40]. In these simulations the value $Q = 0.8$ is used, being obtained as the Monte-Carlo averaging over 100 runs on the tracking interval of 100 s.

If a radar system measures in the SCS coordinates (r, α, β), and tracks in the CCS coordinates, a reasonable choice for the range sensor noise standard deviation, σ_R, is the one percentage of its output at the nominal range, R. Moreover, commonly used standard deviations, σ_α and σ_β, of the angular noises in the azimuth and elevation sensors, respectively, are of the order 1 deg., or 0.017 rad, [40]. By using the transformation from SCS to CCS, one obtains the following observation noise covariance in CCS, [40]

$$R_s = diag\left[\sigma_R^2, \sigma_\alpha^2, \sigma_\beta^2\right]; R_c = TR_sT^T;$$
$$T = \begin{bmatrix} \cos\alpha\cos\beta & -r\sin\alpha\cos\beta & -r\cos\alpha\sin\beta \\ \sin\alpha\cos\beta & -r\cos\alpha\cos\beta & -r\sin\alpha\sin\beta \\ \sin\beta & 0 & r\cos\beta \end{bmatrix} \tag{63}$$

Thus, the position measurement noise variance, in (2), in each of the CCS coordinates is given by the corresponding diagonal component of covariance matrix, R_c, in (63). However, in these simulations, an alternative ad-hoc approach, based on the measurement disturbance representation, in (8) and (9), is used. Assuming the standard deviations, $\sigma_n = 1$ and $\sigma_0 \leq 100\sigma_n$, as well that the observations contain from one to twenty-five outlier percentages, corresponding to the contamination degree from $\delta = 0.01$ to $\delta = 0.25$, one obtains the observation noise variance value $R_k = 400$, as an average of the different values of the free parameters δ and σ_0^2, in (9).

Performances of the next algorithms are compared: optimal KF (3)–(6), designated as A1; adaptive KF, combining the A1 filter with the Huber's moving window M-robust algorithm to estimate noises moments, in (39), (42), (43), (52), (55), (56), (58) and (59), denoted as A2; M-robust estimate KF, in (35)–(38), denoted as A3; and adaptive M-robust estimate KF, combining the filter A3 and the Huber M-robust parameter estimator of unknown noises statistics in A2, denoted as A4. A filter adaptation to the unknown noise statistics requires to adopt the lengths of the corresponding sliding frames. A reasonable ad-hoc choices are $L \in [50, 500]$, $l \in [3, 10]$ and the forgetting factor, in (59), $\gamma \leq 10$. Particularly, the presented simulation results are generated by using $L = 250$, $l = 5$, and $\gamma = 9$. The initial guesses of the noise statistics are chosen as $\hat{r}_1 = 0$, $\hat{R}_1 = 400$, $\hat{q}_0 = 0$ and $\hat{Q}_0 = 0.8$. These constant values of the noise statistics are also used in the non-adaptive filters A1 and A3.

Simulation results, for each of the CCS coordinates, are analyzed using the cumulative estimation error (*CEE*) factor of goodness, together with the estimated observation disturbance moments (mean value and variance), where the criterion

$$CEE(k) = \frac{1}{k}\sum_{i=1}^{k} \frac{\|\hat{x}_i - x_i\|}{\|x_i\|}; k = L, L+1 \tag{64}$$

with $\|\cdot\|$ being the Euclidian norm. Since the obtained results are similar to each of the CCS coordinates, only the results for the z-coordinate are presented in **Figures 4** and **5**, while the estimated observation noise mean value and variance are plotted in **Figure 6**. The commonly used sample variance and mean estimates, generated at moving frame of the length $L = 250$ upon the Gaussian observation noise, are compared with the robustly estimated observation disturbance moments by the algorithm A4 upon the normally distributed observations corrupted by different types of outliers, involving the innovation, observation and structural ones, (**Figure 6**). It should be noted that the adaptive algorithms A2 and A4 provide slightly better results than the algorithm A1, and obviously better results than the algorithm A3 under the pure normally distributed observations in the presence of the target maneuver (**Figure 4**). Moreover, the robust algorithms, A3 and A4, provide better results than the linear non-robust ones, A1 and A2, upon the presence of multiple outliers (**Figure 5**).

Finally, the adaptive robustified algorithm A4 provides the best results, representing an appropriate trade-off between tracking ability and robustness

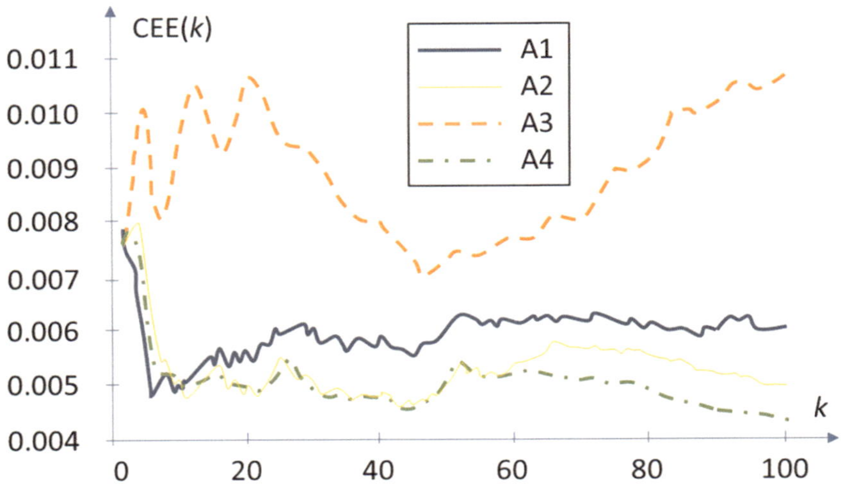

Figure 4.
CEE *criterion comparison of different estimators in the Gaussian observation noise* ($\delta = 0$), *in the presence of a targeting maneuver.*

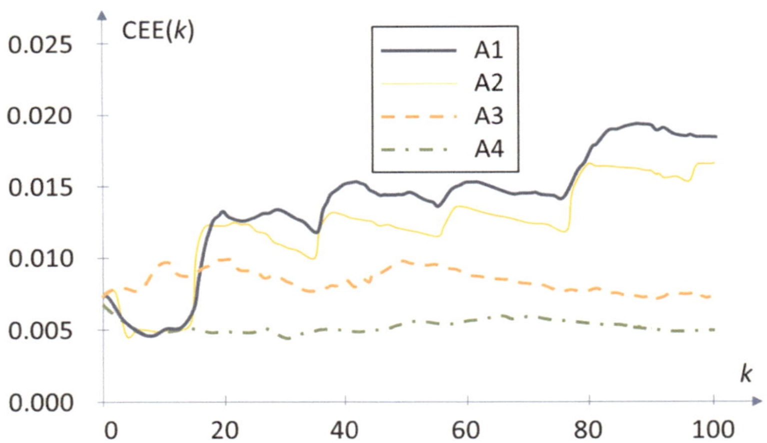

Figure 5.
CEE *criterion comparison of different estimators in the contaminated Gaussian noise* ($\delta = 0.1$), *involving innovation, observation, and structural outliers.*

performance, suppressing efficiently outliers with a contamination degree less than thirty percent, (**Figures 4** and **5**). Also, the estimated noise statistics by the algorithm A4 upon the presence of outliers compare equally with the sample mean and variance estimates upon the absence of outliers, (**Figure 6**).

6. Conclusion

Since the Kalman filter has a computationally attractive recursive structure for real-time applications, it is widely used to solve the practical problems, involving

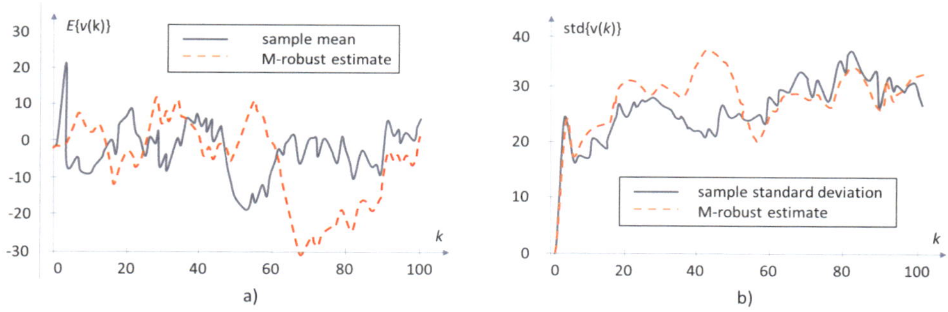

Figure 6.
Estimated measurement noise statistics comparison of sample mean and variance estimates upon Gaussian noise, and the algorithm A4 estimates upon thick-tailed Gaussian noise ($\delta = 0.1$): (a) mean value; (b) standard deviation.

system parameter identification, dynamic system state estimation, optimal and adaptive control, statistical signal processing, etc. In addition, numerous scientific and industrial measurements are analyzed, in the statistical sense, and the obtained results have indicated that the observations contain from five to ten percent of outliers, on average, due to incomplete measurements, meter and communication errors, errors in mathematical model, etc. In this sense, a thick-tailed Gaussian noise distribution is followed by the two types of impulsive noises, inducing innovation and observation outliers. The third type of outliers, named structural outliers, appears due to modeling errors, such as unmodeled system dynamics, time-varying model bias, computational errors, etc. Particularly, the thick-tailed normally distributed observations noise is presented in a maneuvering target tracking, due to the target glint inducing the innovation and observations outliers. Two characteristics of such data can be stressed. Firstly, there exists a high-intensity spiky character of the observations and, secondly, there is a clear low frequency, named the bright spot wonder, being produced by a slow drift in a target gravity center, due to changes in the target aspect. The low frequency can be efficiently suppressed in the tracking loop, but the glint spikes, representing the observation and innovation outliers, are stuck to the tracking loop. Moreover, a linear dynamic state space model is not adequate during a target maneuver, and such modeling errors induce structural outliers. However, since the Kalman filter is a linear signal processor, its performance can be very poor upon the outliers' existence.

The basic objective of this manuscript is to design an appropriate Kalman filter modification that performs fairly well upon multiple outliers. It has been demonstrated, by a real-life maneuvering target tracking example, that the proposed adaptive M-robust estimate-based Kalman filter improves significantly the target estimation and tracking quality, being effective in suppressing multiple outliers with contamination degree less than thirty percent. Furthermore, the achieved improvement is at expense of the additional computational efforts, but usually not so demanded to preclude a real-time application. On contrary, the emphasized approximate Kalman filter may be more efficient than the least squares, or the optimal Kalman filter. Therefore, it may be used to solve the numerous engineering problems in which the Kalman filter, or least squares, is required. Furthermore, it may be also used as an alternative to the other adaptive techniques, involving discrete noise levels, variable state dimension, and interacting multiple model approaches, as well the observation distribution estimation. Further investigations may be directed to propagation of the error covariances upon a non-linear form of the state updates, as well as to the convergence, observability, and consistency analysis of the proposed approximate filter.

Author details

Branko Kovačević[1], Zoran Banjac[2]* and Tomislav Unkašević[2]

1 School of Electrical Engineering, University of Belgrade, Serbia

2 Vlatacom Institute, Belgrade, Serbia

*Address all correspondence to: zoran.banjac@vlatacom.com

References

[1] Gelb A, Kasper J, Nash R, Price C, Sutherland A, editors. Applied Optimal Estimation, Analytic Sciences Corporation. Cambridge, MA: MIT Press; 1974. ISBN 9780262570480

[2] Grewal MS, Andrews AP. Kalman Filtering: Theory and Practice Using Matlab. Hoboken, NJ: Wiley; 2015

[3] Heijden F, van der Lei B, Xu G, Ming F, Zou Y, de Ridder D, et al. Classification, Parameter Estimation, and State Estimation: An Engineering Approach Using Matlab. Hoboken, NJ, USA: John Wiley & Sons, Inc; 2017

[4] Kovačević B, Đurović Ž. Fundamentals of Stochastic Signals, Systems and Estimation Theory with Worked Examples. Berlin: Springer; 2011

[5] Verhaegen M, Verdult V. Filtering and System Identification: A Least Squares Approach. Cambridge: Cambridge University Press; 2012

[6] Price C. An analysis of the divergence problem in the Kalman filter. IEEE Transactions on Automatic Control. 1968;**13**(6):699-702. DOI: 10.1109/tac.1968.1099031

[7] Hanlon P, Maybeck P. Characterization of Kalman filter residuals in the presence of mismodeling. IEEE Transactions on Aerospace and Electronic Systems. 2000; **36**(1):114-131. DOI: 10.1109/7.826316

[8] Albert AE, Gardner LA. Stochastic Approximation and Nonlinear Regression. Cambridge, MA: The MIT Press; 1967

[9] Saridis G, Nikolic Z, Fu K. Stochastic approximation algorithms for system identification, estimation, and decomposition of mixtures. IEEE Transactions on Systems Science and Cybernetics. 1969;**5**(1):8-15. DOI: 10.1109/tssc.1969.300238

[10] Stanković S, Kovačević B. Analysis of robust stochastic approximation algorithms for process identification. Automatica. 1986;**22**(4):483-488. DOI: 10.1016/0005-1098(86)90053-1

[11] Tsypkin, I. Adaptation and Learning in Automatic Systems. New York: Academic Press; 1971

[12] Mendel JM. Adaptive, Learning and Pattern Recognition Systems: Theory and Applications. New York: Acadamic Press; 2012

[13] Masreliez C. Approximate non-Gaussian filtering with linear state and observation relations. IEEE Transactions on Automatic Control. 1975;**20**(1): 107-110. DOI: 10.1109/TAC.1975. 1100882

[14] Masreliez C, Martin R. Robust Bayesian estimation for the linear model and robustifying the Kalman filter. IEEE Transactions on Automatic Control. 1977;**22**(3):361-371. DOI: 10.1109/ TAC.1977.1101538

[15] Myers K, Tapley B. Adaptive sequential estimation with unknown noise statistics. IEEE Transactions on Automatic Control. 1976;**21**(4):520-523. DOI: 10.1109/TAC.1976.1101260

[16] Tsai C, Kurz L. An adaptive robustizing approach to kalman filtering. Automatica. 1983;**19**(3):279-288. DOI: 10.1016/0005-1098(83)90104-8

[17] Banjac Z, Kovačević B. Robustified Kalman filtering using both dynamic

stochastic approximation and M-robust performance index. Tehnicki Vjesnik - Technical Gazette. 2022;**29**(3):907-914. DOI: 10.17559/tv-20200929143934

[18] Pavlović M, Banjac Z, Kovačević B. Approximate Kalman filtering by both M-robustified dynamic stochastic approximation and statistical linearization methods. EURASIP Journal on Advances in Signal Processing. 2023; **2023**(1):69. DOI: 10.1186/s13634-023-01030-1

[19] Barnett V, Lewis T. Outliers in Statistical Data. Chichester: Wiley; 2000

[20] Venables WN, Ripley BD. Modern Applied Statistics with S. New York: Springer; 2011

[21] Huber PJ, Ronchetti EM. Robust Statistics. Hoboken, NJ: Wiley; 2009

[22] Wilcox RR. Introduction to Robust Estimation and Hypothesis Testing. Amsterdam: Academic Press, an imprint of Elsevier; 2017

[23] Hampel FR, Ronchetti EN, Rousseeuw PJ, Stahel WA. Robust Statistics: The Approach Based on Influence Functions. Hoboken, NJ: Wiley; 1986

[24] Dutter R. Numerical solution of robust regression problems: Computational aspects, a comparison. Journal of Statistical Computation and Simulation. 1977;**5**(3):207-238. DOI: 10.1080/00949657708810152

[25] Hogg RV. Statistical robustness: One view of its use in applications today. The American Statistician. 1979;**33**(3):108. DOI: 10.2307/2683810

[26] de Menezes D, Prata D, Secchi A, Pinto J. A review on robust M-estimators for regression analysis. Computers & Chemical Engineering. 2021;**147**:107254. DOI: 10.1016/j.compchemeng.2021.107254

[27] Kovačević B, Milosavljević M, Veinović M, Marković M. Robust Digital Processing of Speech Signals. Berlin Heidelberg: Springer; 2017

[28] Boncelet C, Dickinson B. An approach to robust Kalman filtering. In: The 22nd IEEE Conference on Decision and Control. SA, TX, USA; 1983. pp. 304-305. DOI: 10.1109/cdc.1983.269847

[29] Kovačević B, Đurović Ž, Glavaški S. On robust Kalman filtering. International Journal of Control. 1992; **56**(3):547-562. DOI: 10.1080/00207179208934328

[30] Đurović Ž, Kovačević B. Robust estimation with unknown noise statistics. IEEE Transactions on Automatic Control. 1999;**44**(6): 1292-1296. DOI: 10.1109/9.769393

[31] Chang G, Liu M. M-estimator-based robust Kalman filter for systems with process modelling errors and rank deficient measurement models. Nonlinear Dynamics. 2015;**80**(3): 1431-1449. DOI: 10.1007/s11071-015-1953-0

[32] Gandhi MA, Mili L. Robust Kalman filter based on a generalized maximum-likelihood-type estimator. IEEE Transactions on Signal Processing. 2010; **58**(5):2509-2520. DOI: 10.1109/tsp.2009.2039731

[33] Zou Y, Chan S, Ng T. Robust M-estimate adaptive filtering. IEE Proceedings - Vision, Image, and Signal Processing. 2001;**148**(4):289. DOI: 10.1049/ip-vis:20010316

[34] Banjac ZD, Kovacevic BD, Milosavljevic MM, Veinovic MD. Local

echo canceler with optimal input for true full-duplex speech scrambling system. IEEE Transactions on Signal Processing. 2002;**50**(8):1877-1882. DOI: 10.1109/tsp.2002.800415

[35] Banjac Z, Kovačević B, Veinović M, Milosavljević M. Robust least mean square adaptive FIR filter algorithm. IEE Proceedings - Vision, Image, and Signal Processing. 2001;**148**(5):332-336. DOI: 10.1049/ip-vis:20010594

[36] Kovačević B, Banjac Z, Milosavljević M. Adaptive Digital Filters. Berlin: Springer-Verlag; 2013

[37] Kovačević B, Banjac Z, Kovačević IK. Robust adaptive filtering using recursive weighted least squares with combined scale and variable forgetting factors. EURASIP Journal on Advances in Signal Processing. 2016;**2016**(1):37. DOI: 10.1186/s13634-016-0341-3

[38] Hewer G, Martin R, Zeh J. Robust preprocessing for Kalman filtering of glint noise. IEEE Transactions on Aerospace and Electronic Systems. 1987; **AES-23(1)**:120-128. DOI: 10.1109/TAES.1987.313340

[39] Woolcock SC. Target Characteristics. London: Technical Editing and Reproduction, Ltd; 1973

[40] Blackman SS, Popoli R. Design and Analysis of Modern Tracking Systems. Norwood, MA: Artech House; 1999

Chapter 6

Kalman Filtering Applied to Investment Portfolio Management

Alwyn J. Hoffman

Abstract

Arbitrage is an approach to portfolio composition that times the selection of investment instruments based on deviations between price and true value, exploiting the expected movement of price toward true value. Pairs trading is an early form of statistical arbitrage, exploiting temporary deviations from equilibrium for instruments that tend to move together. To extend the potential investment, universe statistical arbitrage was extended to include several co-integrated instruments that display mean reverting model errors. A weakness of these methods is the possible breakdown of the observed relationships when the model error becomes a random walk, resulting in trading losses. Partial co-integration addresses this limitation by allowing for model errors that include both random walk and mean-reverting components. As both the mean-reverting and random walk components are non-observable, a Kalman filter approach is employed to estimate these system states. Partial co-integration is shown to outperform market returns and normal co-integration. Higher returns are observed during bear cycles, making it an attractive strategy to combine with bull market strategies. It is furthermore illustrated how partial co-integration can be further improved using convolutional neural networks to predict breakdowns in mean reversion and reinforcement learning to optimize the levels for entering and exiting trades.

Keywords: portfolio management, statistical arbitrage, pairs trading, partial co-integration, convolutional neural network, reinforcement learning

1. Introduction

Kalman filtering is a method that enhances the estimation accuracy of uncertain observations by optimally combining the last available measurement with the best prior estimate [1]. The best new estimate is effectively calculated as a weighted average of the previous estimate, considering expected system behavior, and the last measurement. The weighting factors are moderated by the estimated uncertainty of the respective contributions, aiming to minimize the error between the estimate and the true value of the uncertain variable [1].

In the case of portfolio management investment, instruments must be selected to achieve an acceptable balance between risk and return. A key principle of portfolio management is diversification, exploiting the fact that the returns of different instruments are not perfectly correlated [2]. If only long positions are considered, an optimal portfolio will reside at the efficient frontier, which represents the highest return for a given risk, or the lowest risk for a given return [2].

A further improvement can be obtained by also considering short positions, allowing the hedging of risk by assuming a market neutral approach where the portfolio eliminates systematic risk, which is movement synchronous with changes in the overall market of available instruments [2]. In an ideal case, the portfolio will consist of a combination of long and short positions that are based on temporary mispricing observed in the market, with long positions taken in instruments that are underpriced and short positions taken in instruments that are overpriced. During the early stages of trading in financial instruments, it was possible to profit from such trades due to the lack of immediate information from all markets where the same instruments were traded, exploiting temporary differences in prices available in different exchanges. This was called arbitrage, which can be viewed as an approach to portfolio composition that takes long and short positions in the same instrument that is priced differently in different markets, guaranteeing a profit at zero risk [3].

As high-speed communication and information systems eliminated opportunities for this type of arbitrage, investment managers started to search for opportunities where the selection of investment instruments is based on deviations between current price and true value, aiming to exploit the expected movement of price toward true value. True value may be estimated based on observed statistical relationships between investment instruments over time; thus, the term statistical arbitrage was coined for a strategy that exploits deviations between current price and true value derived from long-term statistical relationships [4]. Pairs trading is a form of statistical arbitrage that involves the selection of two financial instruments that tend to move together [5]. For such instruments, when one is regressed in terms of the other, the model error tends to be mean reverting. An instrument that is known to be mean reverting can be exploited for profit: when the model error deviates beyond a set limit, a long position is taken in the instrument that is currently priced too low, given the price of the other, and a short position is taken in the instrument that is currently priced too high, based on the expectation that the model error will revert to its mean zero value [5].

Statistical arbitrage can be extended to include several co-integrated instruments. Rather than selecting only two co-integrated instruments, a group of co-integrated instruments is identified [6]. If the group is co-integrated and one of the instruments is regressed in terms of the others, the model error will be stationary and thus mean-reverting. If this model errors moves below a specified negative threshold, it is expected to return to its mean value of zero within a limited period, vice versa when the model error exceeds a positive threshold. Depending on the sizes and signs of the regression model coefficients, positions can be taken in each of the instruments so that the portfolio value will equal the model error [6]. Movement of the model error from a negative value back to zero will result in an overall profit for the portfolio. If opposite positions are taken for a movement of the model error from a positive value back to zero, a profit will once again be generated by the portfolio. Higher returns are observed during bear cycles, making statistical arbitrage an attractive strategy to combine with bull market strategies [7].

A weakness of this technique is the fact that the observed relationships may break down, causing the model error to behave like a random walk [7]. This may lead to large losses if the model error, after exceeding the predefined threshold, does not revert to the mean but continues a random walk away from zero until a stop loss threshold is exceeded. To counter this phenomenon, the concept of partial co-integration was introduced [7]. Partial co-integration allows for model errors that include both random walk and mean-reverting components. A Kalman filter approach is used to extract the mean-reverting and random walk components that cannot be observed directly. Trading signals are based on the mean-reverting component only, resulting in more frequent trades and thus more profit opportunities. Improvement of risk-adjusted returns generated by the partial co-integration strategy is enabled by the optimal selection of the Kalman filter gain [8]. Such a strategy can be shown to outperform market returns and normal co-integration.

The random walk component, which represents part of the returns that are generated when trades are executed, may however cause losses that will reduce overall portfolio profitability. A further improvement on the above approach is to predict the probability of a breakdown of the co-integrated relationship [9]. To generate information-rich features based on which changes in co-integrated behavior may be detected at an early stage, it is necessary to consider both time and frequency behavior of the price time series. Continuous wavelet transforms may be used for this purpose, transforming one-dimensional time data into a two-dimensional time-frequency space [9]. Convolutional neural networks are then applied to detect changes in the 2D behavior of the price series.

The portfolio optimization process can be based on reinforcement learning, which regards the portfolio as a Markov decision process [9]. Each possible portfolio composition is regarded as a system state. For each state, any of several actions can be taken to change the system to a different state; this represents new positions taken in the underlying instruments. For each state and action combination, a consequence is registered, this is, the profit or loss that results over the next period. By learning from past behavior through reinforcement learning, a model can be developed that makes an optimal decision for each possible system state. The objective function of the optimization process takes into account the probability that the co-integrated relationship may break down [9].

In this chapter, we review and compare the above approaches toward statistical arbitrage. It is shown that Kalman filters and partial co-integration can be successfully combined with deep learning and reinforcement learning to address the weakness observed in more traditional approaches to statistical arbitrage.

2. Statistical arbitrage and pairs trading

Arbitrage pricing theory (APT) was first suggested by Ross [3] to determine asset prices. Statistical arbitrage encompasses quantitative trading strategies with the following features: "(i) trading signals are systematic, or rules-based, as opposed to driven by fundamentals, (ii) the trading book is market-neutral, in the sense that it has zero beta with the market, and (iii) the mechanism for generating excess returns is statistical" [10]. Using market neutral strategies or long/short strategies to produce low-volatility investment strategies that take advantage of diversification across assets is very popular in the hedge fund industry [11].

Many statistical arbitrage strategies are focused on the concept of mean-reversion of security prices. The concept of mean-reversion trading is built on the assumption that a security's high and low prices are only temporary and that the series will revert to a certain mean value over time [12]. While mean-reversion strategies are effective when price series are stationary, the price series of most securities are not stationary due to drifts caused by trends and momentum.

Nonstationary time series can be converted to stationary time series by differencing. To obtain a covariance stationary series, the series must be differenced by a minimum order of integration d, denoted $I(d)$. While single price series are seldom stationary, it is possible to obtain a stationary price series by creating a linear (weighted) combination of securities that exhibit a co-integrated relation. If the returns of two or more series are individually integrated of the same order and a linear combination of the series is integrated of lower order, the series are said to be co-integrated [13]. Some co-integration testing techniques include the Engle-Granger two-step method [14], the Johansen test [6], and the Phillips-Ouliaris test [15]. In contrast to the Engle-Granger method and Phillips-Ouliaris test, the Johansen test can be used to test multiple time series for co-integration.

Some forms of statistical arbitrage are pairs trading [16] and long/short strategies [17]. Pairs trading has been known in the quantitative finance community since the Mid-1980s [5]. It involves identifying two securities whose prices tend to travel together. The difference in prices is called the spread. Upon divergence of the spread beyond a defined threshold, the cheaper security is bought long and the more expensive one is sold short. When the prices converge back to their historical equilibrium, the trade is closed, and a profit is collected. The formalization of these concepts results in a quantitative trading strategy that exploits relative mispricing between two securities. Pairs trading can thus be regarded as a relative-value arbitrage strategy.

Zou et al. [18] pioneered pairs trading as a popular market-neutral trading strategy to exploit co-integrated relations that exist in the market between pairs of instruments. Vidyamurthy [19] suggested that APT can be used to detect tradeable pairs. Gatev et al. identified pairs to which statistical arbitrage could be applied using a minimum distance method to [16]. After obtaining securities that have a co-integrated relation, a certain hedge ratio must be determined for the pair; different approaches to calculate this hedge ratio have been investigated. Previous studies [20, 21] provide promising results employing the Kalman filter to determine the hedge ratio dynamically.

Gatev et al. [16] applied pairs trading to US CRSP securities for the period 1962 to 2002. They used sum of squared distances in normalized price space to rank all pairs in a 12-month formation period. They transferred the top 20 pairs with minimum distance metric to a subsequent 6-month trading period, entered a trade when the spread diverges at least two historical standard deviations from equilibrium, and closed it with the next zero-crossing, at the end of the trading period, or upon delisting. Using this approach, they achieved statistically and economically significant excess returns of 11% p.a., exhibiting low exposure to common sources of systematic risk.

3. Co-integration trading strategy

Pairs trading is profitable only if the spread between the securities is mean reverting, ensuring that it will return to zero within a reasonable period after diverging beyond the trading threshold. This requirement can be described in terms of the concept of co-integration. Two variables (in our case the prices of two instruments) are co-integrated if the error term resulting from regressing one in terms of the other is stationary [13]. Johansen extended this to an arbitrary number of variables involved in the regression [6]. A time series can be tested for stationarity using the Augmented Dickey-Fuller (ADF) [22] test. If a time series passes the test and is indeed stationary, a dependency is expected to exist between historic and future values: if previous values were above the mean, it is expected that upcoming values will move down toward the mean and *vice versa*.

The error term for a co-integrated set of variables can be described using an error-correction model or ECM [23]. The dynamics of one time series at a certain time point can then be described as a correction of the last period's deviation from the equilibrium with the addition of possible lag dynamics. The advantage of using an ECM is that active predictions can be simply done by using past information.

To test for co-integration, the two-step procedure proposed by Engle and Granger [13] must be applied. First, a linear regression is run of the one series on the other. If the input and output variables x_t and y_t are co-integrated of order one, the error term ε_t of the regression will be weakly stationary (i.e., integrated of order zero). This is described by the following equation:

$$y_t = c_0 + bx_t + \varepsilon_t \tag{1}$$

Based on the work of Johansen [6], x_t can be extended to a vector containing an arbitrary number of variables. The composition of the co-integration portfolio is dictated by the regression coefficients of the model. A portfolio can be obtained for which the combined value at any point in time corresponds to the model error or residual signal by investing into each instrument in proportion to the size of the regression coefficient.

A trading strategy must then be defined for opening and closing positions. The residual signal of each co-integrated set must be checked dynamically against a z-score that is calculated using the moving average μ_ε and standard deviation σ_ε of the errors that were determined during the formation period:

$$z_t = \frac{\varepsilon_t - \mu_\varepsilon}{\sigma_\varepsilon} \tag{2}$$

Trading signals are generated when z_t exceeds predetermined thresholds; for instance, when $z_t = -2$, a long position is taken in the portfolio, while a short position is taken when $z_t = 2$. The trade is closed when the spread has closed to a value of zero; alternative closing objectives may also be used.

Shares must be available for significant periods to extract reliable relationships based on which they can be tested for co-integration. A share must be available for the complete co-integration period to be considered for inclusion in a co-integrated set. For each group of instruments, a new model is extracted and tested for a sufficient level of co-integration on a periodic basis, using the h-values of the Engle-Granger or Johansen test. The resulting model error is tested for stationarity by applying the Dickey-Fuller test. A model is only accepted to compile a co-integrated portfolio if both tests are passed. In such a case, the same model is used until the conditions are satisfied for exiting the trade, that is, once the model error reaches a zero value. The periodic compilation of new models is then resumed until a new co-integrated relationship is found. The entry condition based on the z-value of Eq. (2) may be varied as part of the optimization process.

In the case of short positions, the negative values of the model coefficients are used to determine the effective positions in the shares. As model coefficients can have any sign, in most cases, portfolios consist of both short and long positions in the various instruments. As a result, the effective net investment position is mostly close to zero, given that long positions are partly offset by short positions in other shares.

A typical trade that is entered and exited is illustrated in **Figure 1** below. The closing prices of the shares forming part of the co-integrated set are displayed in the first graph. The second graph displays the error signal and the high and low thresholds that must be exceeded to trigger a trade. These thresholds vary over time as the level of volatility of the error signal changes over time. This is of specific relevance as the size of the error at the point of entering a position depends on the standard deviation of the error at that stage. The volatility of the underlying shares and thus of the model error is changing all the time, thus leading to widely varying standard deviations. This phenomenon is called heteroscedasticity and is often observed among financial instruments.

The position where a long trade is entered, once the error signal breaks through the upper threshold, is indicated by a blue star, and the red star indicates the point where the trade exits once the error signal breaks downwards through zero. The closing price graph displays the points where the trades occurred, as well as the returns produced by each share. The net return of the trade, the stop loss, and profit target levels as functions of time are displayed in the bottom graph—in this case, a loss was registered

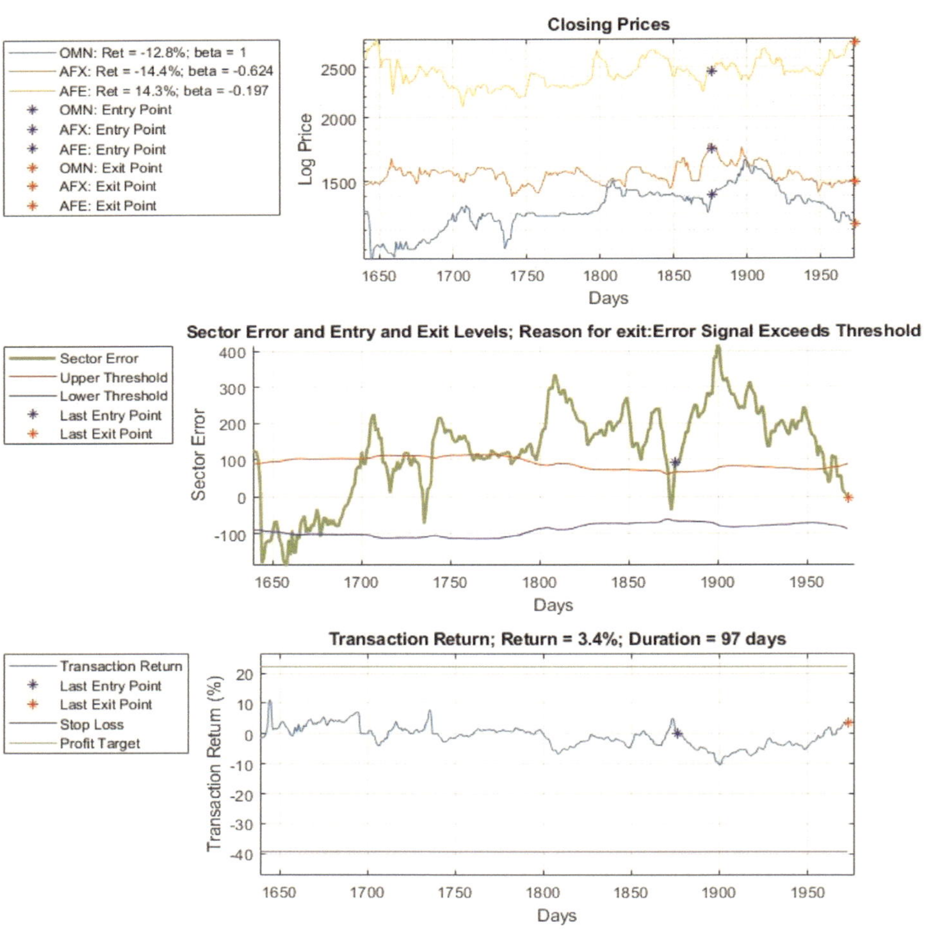

Figure 1.
*Illustration of a typical co-integration strategy trade. Graph 1: Closing prices of a group of co-integrated shares. Graph 2: Error signal and trading thresholds, with entry and exit point indicated by *. Graph 3: Net return, stop loss, and profit target.*

after entering the trade, as the error signal moved further above the threshold, but this turned into a profit by the time that the trade exits.

If the entire model error is used to trigger trades, entering at the threshold and exiting at zero, a guaranteed profit is made (excluding trading costs) each time that a position is taken and exited. An additional condition is however that the model error must indeed reverts to zero within an acceptable period. The model error could move away from zero indefinitely, should the co-integrated relationship between the instruments break down after a position was taken, as is typical for random walks. An unlimited loss can then be incurred on the position that was assumed, unless stop losses are implemented.

Another possibility is that a profit is shown before the zero-error level is reached, but that this turns into a loss before satisfying the zero-crossing condition for exiting the trade. A profit target may be set to exploit such profits by providing an alternative condition to exit the strategy on a profitable basis.

Clegg [24] found that for all shares forming part of the S&P 500 on the NYSE, the average proportion of variance attributable to mean reversion was 37%; thus, about 19% of price fluctuations in a given day would be attributable to the mean-reverting component. The average mean reversion component ρ was only 0.06, with ρ defined as the AR(1) coefficient of the mean-reverting price signal p_{mr} at time t and $\varepsilon_{mrs,t}$ the model error:

$$p_{mr,t} = \rho p_{mr,t-1} + \varepsilon_{mrs,t} \tag{3}$$

Both the proportion of variance attributable to mean reversion and the probability of persistence of mean reversion increase with ρ.

The model error (or spread) for a set of shares that satisfied the conditions for co-integration at the time of model extraction often tends to display behavior that is a combination of a random walk and mean reversion [8]. The random walk component of the error signal reduces the frequency of mean reversion, as it prevents the spread from reverting back to zero and may in some cases prevent the error signal from returning to zero at all, causing stop loss exits that erode profitability. These limitations suggested the need for a method that can differentiate between the mean-reverting and random walk components of the spread and thus the extension of the concept of co-integration to include partial co-integration.

4. Partial co-integration and the application of Kalman filtering

The weakness of trading strategies based on normal co-integration is the fact that the spread (or model error) is most often not mean reverting, but rather a combination of a random walk and mean-reverting behavior. The spread can be directly observed, using the respective share prices and the values of the extracted regression coefficients. The different error signal components contributing to the spread are however not directly observable.

Clegg and Krauss [7] proposed a solution that extends the model in Eq. (1) above as used by Engle and Granger. They did this by providing for an error that consists of a random walk element, a mean-reverting autoregressive element, as well as a stochastic component. As the different error signal components are not directly observable,

Clegg and Krauss restated the model in state space to allow the unknown variables to be estimated using a Kalman filter approach.

A Kalman filter estimates a value for a stochastic variable that cannot be measured directly by using known relationships between observable variables and so-called system states that cannot be observed directly. In this case, the different states to be estimated from observations are the mean-reverting error component $\varepsilon_{mr,t}$ and the random walk error component $\varepsilon_{rw,t}$. Eq. (1) is thus restated as follows:

$$y_t = c_0 + bx_t + \varepsilon_t = c_0 + bx_t + \varepsilon_{s,t} + \varepsilon_{mr,t} + \varepsilon_{rw,t} \tag{4}$$

The spread can therefore be written as:

$$\varepsilon_t = y_t - c_0 - bx_t = \varepsilon_{s,t} + \varepsilon_{mr,t} + \varepsilon_{rw,t} \tag{5}$$

where the mean-reverting and random walk components of the spread can be written as:

$$\varepsilon_{mr,t} = \rho\varepsilon_{mr,t-1} + \varepsilon_{mrs,t} \tag{6}$$

$$\varepsilon_{rw,t} = \varepsilon_{rw,t-1} + \varepsilon_{rws,t} \tag{7}$$

where it is assumed that the noise components of the system states are independent stochastic processes with zero means:

$$\varepsilon_{mrs,t} \sim N\left(0, \sigma_{mr}^2\right) \tag{8}$$

$$\varepsilon_{rws,t} \sim N\left(0, \sigma_{rw}^2\right) \tag{9}$$

For stochastic variables that cannot be measured accurately, a Kalman filter effectively estimates an accurate value by calculating a weighted average between the value estimated from previous observations and the value of the current observation, which is assumed to be uncertain [1]. The relative weights for the two contributions to the current estimation are based on the levels of uncertainty that exists about true values of the two contributions, derived from historic noise figures.

In general, the state space representation consists of two equations, a state equation and an observation equation, which may respectively be written as follows [24]:

$$Z_t = F_{t-1}Z_{t-1} + G_{t-1}U_{t-1} + W_{t-1} \tag{10}$$

$$X_t = H_tZ_t + V_t \tag{11}$$

where Z_t is the state of the system that is not be directly observable (in this case the mean reverting and random walk contributions to the spread), while X_t represents the observable parameters, in this case the share prices and resulting total spread. U_t is a possible control input, which in our case is assumed to be zero. F_{t-1} describes how the system changes from one sample period to the next in case of no external inputs U_{t-1}, while W_t and V_t are noise terms with covariance matrices Q_t and R_t.

The Kalman filter algorithm makes it possible to determine the optimal estimate of the hidden states Z_t based on previous observations and the assumed values of the system parameters. The Kalman filter equations are given as follows [24]:

$$P_t^- = F_{t-1}P_{t-1}^+F_{t-1}^T + Q_{t-1} \tag{12}$$

$$P_t^+ = (I - K_t H_t) P_t^- (I - K_t H_t)^T + K_t R_t K_t^T \tag{13}$$

$$K_t = P_t^- H_t^T \left(H_t P_t^- H_t^T + R_t \right)^{-1} \tag{14}$$

$$Z_t^- = F_{t-1} Z_{t-1}^+ + G_{t-1} U_{t-1} \tag{15}$$

$$Z_t^+ = Z_t^- + K_t \left(X_t - H_t Z_t^- \right) = Z_t^- (1 - K_t H_t) + K_t X_t \tag{16}$$

A crucial component of these equations is the Kalman gain matrix K_t, as it determines the influence that a new observation has upon the new estimate of the hidden states Z_t. This is clearly displayed in Eq. (15) above: the new system state Z_t^+ after a new observation X_t is a linear combination of the state before a new observation Z_t^- and the new observation X_t. For a very small Kalman gain (which represents the case where new observations are regarded as very uncertain), X_t has a small weight, and thus, the new state estimation is almost the same as the previous state estimation, while for a high Kalman gain (which represents the case where there is a lot of confidence in new observations), the new state estimate is mainly determined by the new observation.

In the case of partial co-integration, the observable variables are the prices of the shares, while the different states to be estimated from observations are the mean-reverting error component $\varepsilon_{mr,t}$ and the random walk error component $\varepsilon_{rw,t}$. We write the observation equation and state equation, respectively, as follows, similar to the equations used by Clegg and Krauss [7]:

$$X_t = \begin{bmatrix} y_t \\ x_t \end{bmatrix} = \begin{vmatrix} b & 1 & 1 \\ 1 & 0 & 0 \end{vmatrix} \begin{vmatrix} x_t \\ \varepsilon_{mr,t} \\ \varepsilon_{rw,t} \end{vmatrix} \tag{17}$$

$$Z_t = \begin{vmatrix} x_t \\ \varepsilon_{mr,t} \\ \varepsilon_{rw,t} \end{vmatrix} = \begin{vmatrix} 1 & 0 & 0 \\ 0 & \rho & 0 \\ 0 & 0 & 1 \end{vmatrix} \begin{vmatrix} x_{t-1} \\ \varepsilon_{mr,t-1} \\ \varepsilon_{rw,t-1} \end{vmatrix} + \begin{vmatrix} \varepsilon_{xs,t-1} \\ \varepsilon_{mrs,t-1} \\ \varepsilon_{rws,t-1} \end{vmatrix} \tag{18}$$

where $\varepsilon_{xs,t-1}$, $\varepsilon_{mrs,t-1}$, and $\varepsilon_{rws,t-1}$ are the innovations of the state variables x_t, $\varepsilon_{mr,t}$ and $\varepsilon_{rw,t-1}$. The unknown parameters are the mean reversion parameter $-1 < \rho < 1$ and the variances of the mean-reverting and random walk processes $\sigma_{mr}{}^2$ and $\sigma_{rw}{}^2$. In the case of a steady state system, the unknown parameters are constants. These equations can be solved to estimate the values of the mean-reverting error and random walk error. Assuming that the mean reversion and random walk processes are independent, we can also define the proportion of variance attributable to mean reversion as [24]:

$$R_{mr}^2 = \frac{2\sigma_{\mathrm{mr}}^2}{2\sigma_{\mathrm{mr}}^2 + (1+\rho)\sigma_{\mathrm{rw}}^2} \tag{19}$$

It is then possible to solve for the unknowns as follows [24]:

$$\rho = -\frac{v_1 - 2v_2 + v_3}{2v_1 - v_2} \tag{20}$$

$$\sigma_{\mathrm{mr}}^2 = \frac{1}{2}\left(\frac{\rho+1}{\rho-1}\right)(v_2 - 2v_1) \tag{21}$$

$$\sigma_{\mathrm{rw}}^2 = \frac{1}{2}\left(v_2 - 2\sigma_{\mathrm{mr}}^2\right) \tag{22}$$

where

$$v_k = \frac{\left(\rho^k - 1\right)^2 + \left(1 - \rho^{2k}\right)}{1 - \rho^2}\sigma_{\mathrm{mr}}^2 + k\sigma_{\mathrm{rw}}^2 \tag{23}$$

The condition $2v_1 - v_2 = 0$ is satisfied when $\rho = 1$, and as $\rho < 1$, the equations are well-defined. Each partial autoregressive series therefore has a unique parameterization. The above results can be used to obtain the following steady state solution for the Kalman gain matrix [24]:

$$K = \begin{pmatrix} \dfrac{2\sigma_{\mathrm{mr}}^2}{\sigma_{rw}\left(\sqrt{(\rho+1)^2\sigma_{\mathrm{rw}}^2 + 4\sigma_{\mathrm{mr}}^2} + \rho\sigma_{rw} + \sigma_{rw}\right) + 2\sigma_{\mathrm{mr}}^2} \\ \dfrac{2\sigma_{rw}}{\sqrt{(\rho+1)^2\sigma_{\mathrm{rw}}^2 + 4\sigma_{\mathrm{mr}}^2} - \rho\sigma_{rw} + \sigma_{rw}} \end{pmatrix} \tag{24}$$

From the above equations, the unknown state variables can be determined as explained in the next section.

5. Partial co-integration trading strategy

After extracting the regression coefficients between the respective instruments and verifying for cointegration, the Kalman filter parameters are estimated using the equations provided in the previous section. This also allows the mean-reverting and random walk components of the spread to be estimated, as follows: the algorithm calculates the Kalman gain using ρ, σ_{mr} and σ_{rw}. Initially, it is assumed that $\varepsilon_{mr,t} = 0$ and $\varepsilon_{rw,t} = \varepsilon_{s,t}$, that is, the spread, is regarded as a random walk. The algorithm then iterates through the remaining observations X_t. For each observation, the Kalman algorithm equations are used to produce the hidden state Z_t, that is, $\varepsilon_{mr,t}$ and $\varepsilon_{rw,t}$ at that time.

Only the mean-reverting error component is used to trigger trades, rather than trading on the total error; not only will this increase the trading frequency—it will also prevent situations where indefinite losses are incurred due to long-term drift in the random walk component.

Hoffman [8] investigated the relationship between the values of ρ and K_t and the profitability of the partial co-integration strategy. He extracted the value of the regression coefficients β by using the techniques of Engle-Granger or Johansen and then optimized the profitability of the trading strategy profit in terms of the values of ρ and K_k. This enabled significantly improved net trading returns, as it allows the fraction of the error signal residing in the mean-reverting components to be controlled. He found that the value of the mean-reverting coefficient ρ determines the fraction of the total co-integration error that resides in the random walk and the mean-reverting components, respectively; this is in line with the results of Clegg and Krauss [7].

As the underlying behavior of the instruments and the nature of their relationships tend to change over time, the following set of parameters should be optimized from time to time [8]:

1. The threshold for entering trades, defined as a multiple of error standard deviations.
2. The level of the stop loss, also defined as a multiple of error standard deviations.
3. The period over which the standard deviation of the error is measured.
4. Time window used to extract the co-integration relationships.
5. The level α of statistical significance for performing co-integration tests.
6. The profit target level.
7. Mean reversion coefficient ρ.
8. Kalman filter gain.

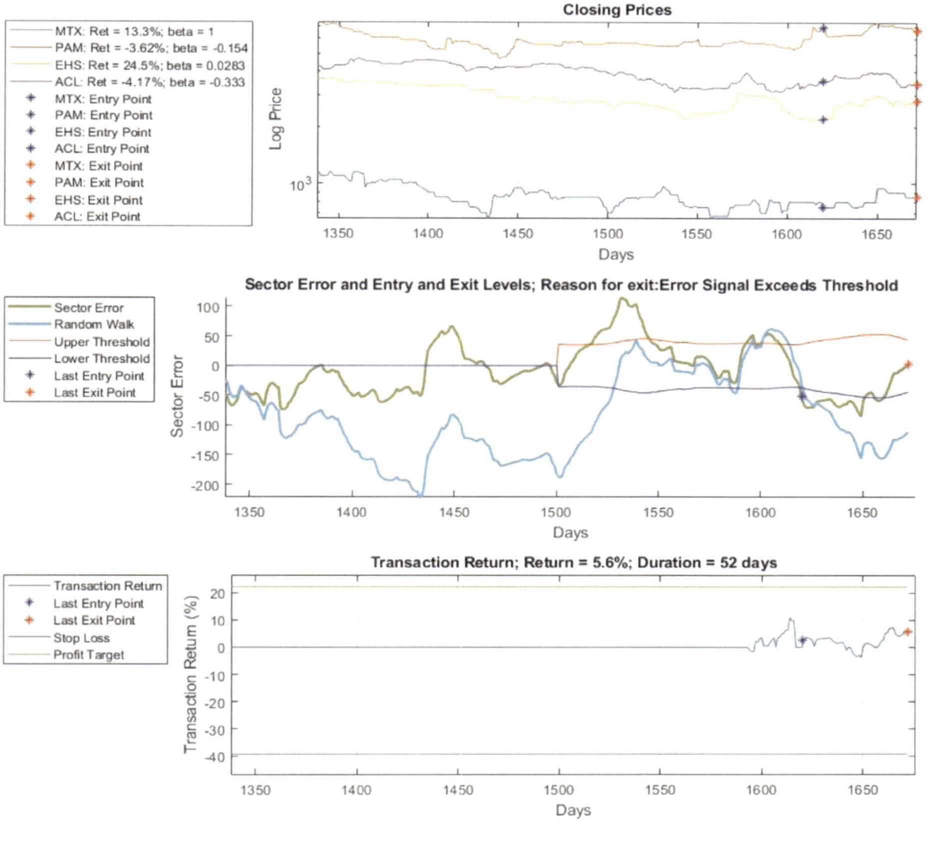

Figure 2.
Illustration of a typical partial co-integration strategy trade. Graph 1: Closing prices of a group of co-integrated shares. Graph 2: Mean reverting and random walk error signals and the trading thresholds, with entry and exit point indicated by X. Graph 3: Net return, stop loss, and profit target.

9. The minimum allowed value for the ratio between the standard deviation of the mean-reverting error signal and the standard deviation of the random walk error signal.

Figure 2 illustrates a typical trade that is based on partial co-integration. In the second graph, the error signal now consists of a mean-reverting and a random walk component; this is the main difference with the co-integration case in **Figure 1**. The trade is entered as the mean-reverting component breaks downward through the lower trading threshold and exits when this component returns to zero. The net return is the sum of the movements of both the mean-reverting and the random walk components. In this case, the trading profit was reduced as the random walk component performed a movement contrary to the movement of the mean-reverting component.

Hoffman [8] implemented a partial co-integration trading strategy on JSE shares by repeatedly extracting co-integrating share sets from 18 industry sectors for consecutive time periods. The available funds were divided equally between available trades from each of the 18 sector strategies to allow maximum diversification between all the available shares. This was repeated after each period where trading positions were assumed. Some sectors produced very few trades, while other sectors provided almost continuous trading opportunities.

In line with Mashele et al. [25], he assumed a trading cost equal to 0.2% of the size of the trade upon entry and exit, resulting in a 0.4% total trading cost per trade. As statistical arbitrage is a high trading strategy, he also investigated the relationship

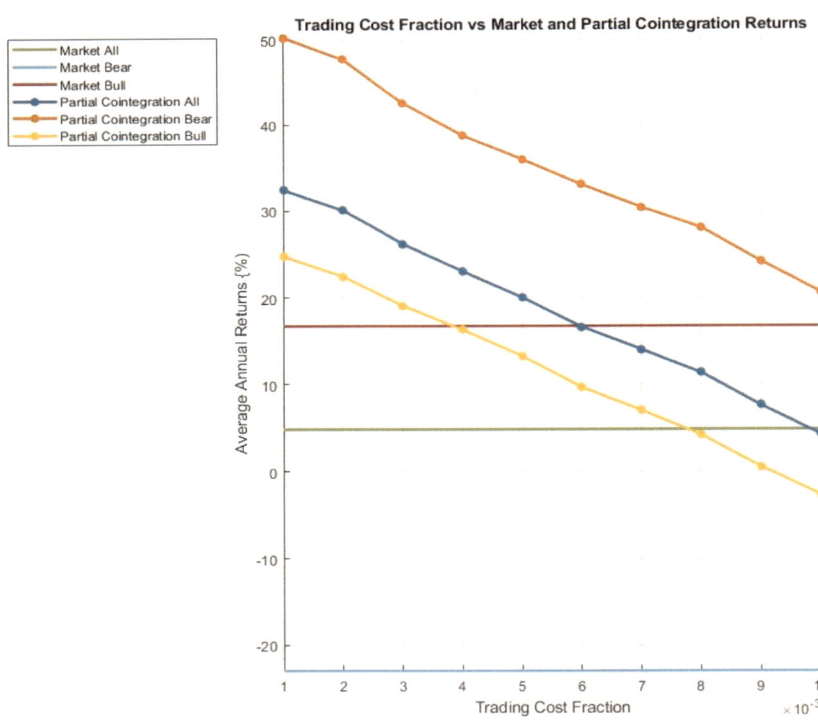

Figure 3.
Average annual profits of the partial co-integration trading strategy on the JSE over the period 1996–2020 for different values of transaction cost.

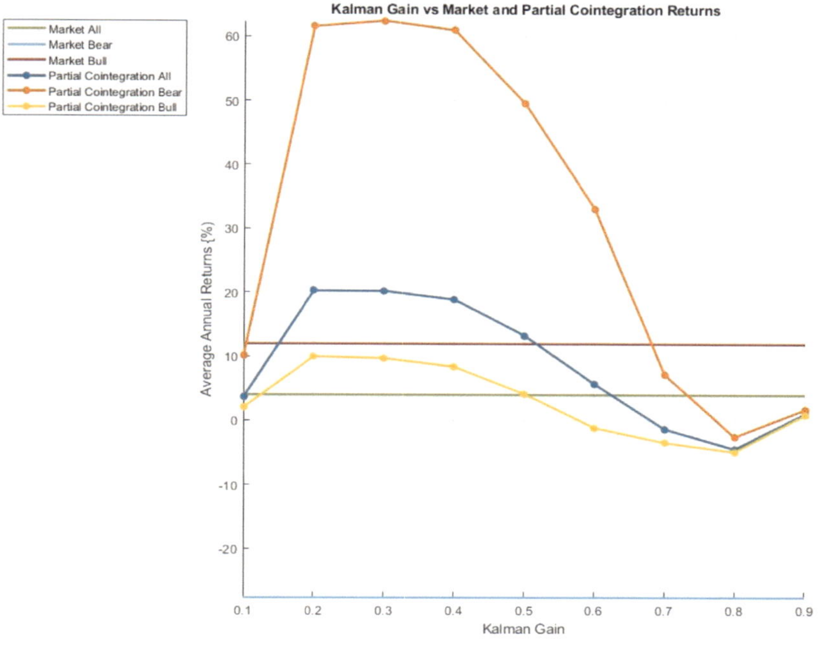

Figure 4.
Average annual profits of the partial co-integration trading strategy over the period 1996–2020 for different values of Kalman filter gain.

between transaction cost and trading profits. The results are displayed in **Figure 3**. The expected reduction in profits can be observed as trading cost is varied from 0% to 1% per complete trade. A profit is still realized when transaction cost assumes its expected value.

Figure 4 shows that the Kalman filter gain significantly influences trading profitability, with an optimal value of 0.7. Hoffman [8] also investigated the impact of trading threshold on performance, as the co-integration error signal levels where trades are initiated dictate the trading strategy. The trading parameter in this case is the value of z as reflected in Eq. (2) above, that is, the multiplier used to determine the error threshold in terms of the mean-reverting error signal standard deviation. As is clear from **Figure 5**, a value of 1.25 was close to optimal. A smaller number of trading opportunities exists for larger values, while the full profit potential of mean-reverting swings of the error signal is not exploited for smaller values. Repeating the calculations for the first and second halves of the available data set, respectively, indicated that the optimal values for the trading threshold are very similar for both data sets.

The trading strategy was implemented for the entire available date range. Subsequently, the annual profits (or losses) generated by the strategy as well as the average annual returns over the entire period were calculated. **Figure 6** compares the accumulated returns generated over the period under investigation by the partial co-integration strategy against the market index and the co-integration trading strategy. The partial co-integration strategy far outperformed the market returns and normal co-integration returns. In addition, the returns generated by partial co-integration were more stable over time compared to market returns.

To confirm this finding, Hoffman [8] also calculated the risk adjusted profits, using the Sharpe ratio [26]. The results are displayed in **Table 1** [8].

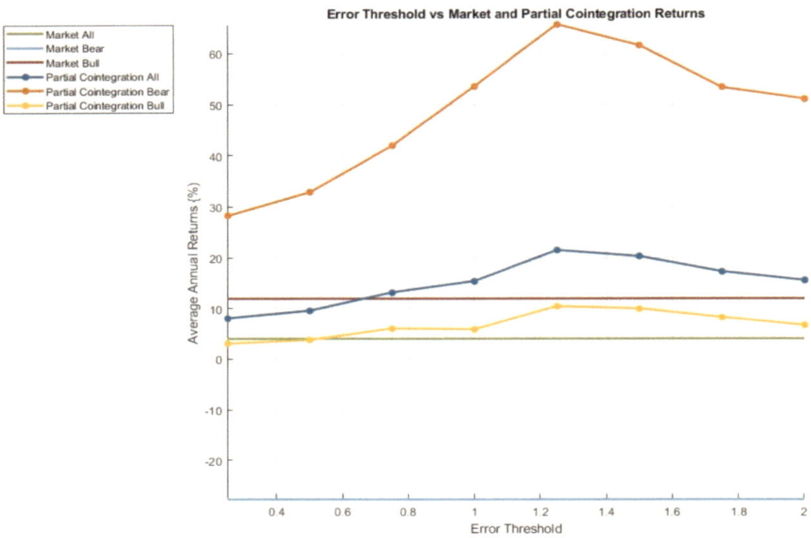

Figure 5.
Average annual profits of the partial co-integration trading strategy over the period 1996–2020 for different values of trading threshold multiplier.

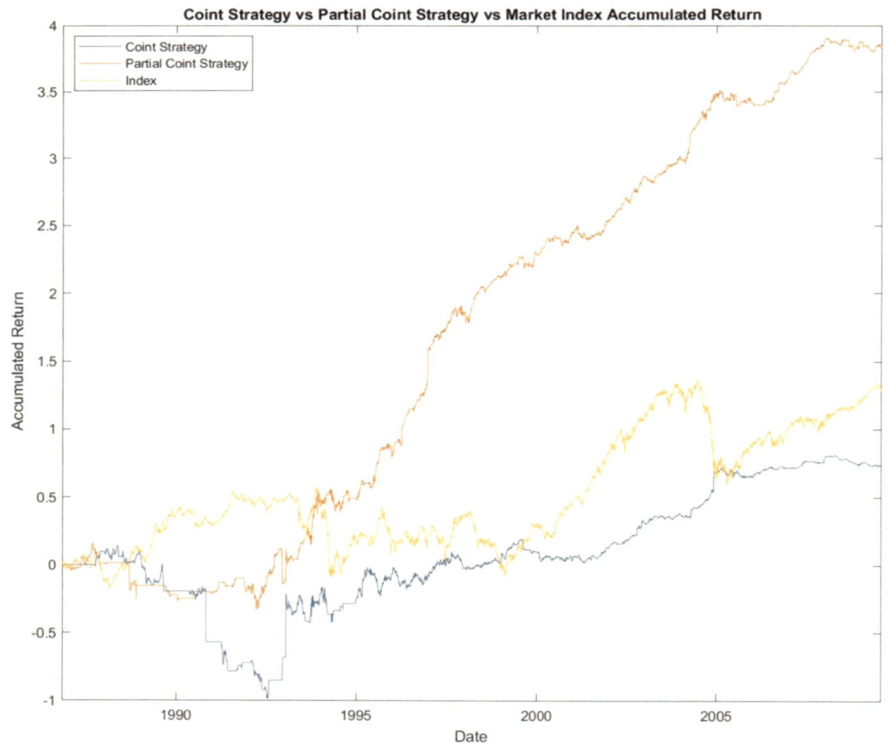

Figure 6.
Comparison of the accumulated returns of the market index with the co-integration and partial co-integration strategies.

	Market	Co-integration	Partial co-integration
All	4.96%	3.81%	22.32%
Low market return	−10.85%	5.78%	27.87%
High market return	19.71%	1.84%	17.15%
Bear cycles	−13.26%	12.54%	29.79%
Bull cycles	17.82%	−1.84%	17.05%
Sharpe ratio	0.15	0.09	0.87

Table 1.
Average annual returns during different parts of the market cycle [8].

The partial co-integration strategy achieved an overall annualized returns of 22.3% compared to market returns of 4.96%. The Sharpe ratio, which is a generally accepted measure of risk-adjusted returns, was 0.87 for partial co-integration compared to 0.15 for market returns. Over the evaluation period of more than 25 years, partial co-integration therefore significantly improved average returns, and reduced the volatility of returns. The normal co-integration strategy could not outperform the market in terms of either average returns or risk-adjusted returns, confirming that partial co-integration produces superior results compared to normal co-integration and outperforms market returns on a risk-adjusted basis.

It has been reported before [7] that statistical arbitrage tends to perform better during bear compared to bull markets. The performance of the co-integration and partial co-integration strategies were therefore separately evaluated for bear and bull markets on the JSE for the available period [8]. Bear and bull cycles were defined as years where market returns were, respectively, below or above a risk-free rate of return set at 2%. The average annual results are separately displayed in **Table 1** for all market cycles, for the lower and upper halves of trading years sorted based on annual returns and for bear and bull cycles [8].

While the partial co-integration strategy performs similar to the market during bull and above average years, it outperforms the market by approximately 40% in terms of annualized returns during bear cycles and below average years. In contrast, the normal co-integration strategy outperforms the market during bear cycles and below average years but performs poorly during bull cycles and above average years [8].

The reasons for exiting co-integration and partial co-integration trading positions are displayed in **Table 2**. The partial co-integration strategy mostly exits positions because the mean-reversion error exceeded its threshold value, indicating that the trading strategy behaves as intended. Positions are exited because the profit target has been reached in about 25% of the cases, while very few positions exit because the stop loss was reached. In contrast, the co-integration strategy hits stop losses more often.

Number of exits/ reason	No co-integration	Reached profit target	Reached stop loss	Same shares not available	Error signal exceeds threshold
Co-integration	0	395	34	13	17
Partial co-integration	0	527	10	7	1636

Table 2.
Reason for exiting trading positions [8].

The total number of trades is also much higher for partial co-integration compared to that for normal co-integration.

Clegg and Krauss [7] provide results obtained for S&P 500 shares (representing about 80% of NYSE market capitalization) over the period 1990 to 2015, which is representative of mature markets that are expected to be more efficient than developing markets like the JSE. They found that compared to average annualized market returns of 9.6%, co-integration produced only 1.09% after transaction costs, while partial co-integration managed to beat the market with 12.34%. Partial co-integration also displayed a lower standard deviation than the market (8.24% compared to 14.99%). Due to its low average returns, co-integration achieved a Sharpe ratio of −0.37, while partial co-integration achieved 1.11 compared to a market value of 0.43. It would therefore appear that partial co-integration offers a viable statistical arbitrage trading strategy.

Clegg and Krauss [7] also compared partial co-integration with standard co-integration and with the market for different periods:

- Jan 1990–Mar 2001 (before the dotcom bubble);
- Apr 2001–Aug 2008 (after the dotcom bubble and before the global financial crisis);
- Sep 2008–Dec 2009 (during the global financial crisis);
- Jan 2010–Oct 2015 (after the global financial crisis).

They found that the excess returns achieved by partial co-integration were severely eroded since the global financial crisis. While the market achieved annualized returns of about 13% for both the first and last periods above, the returns on partial co-integration deteriorated from 22.3% to −2.76%. This is an indication that on mature markets, in contrast with developing markets, the use of sophisticated techniques has increased market efficiencies to a level where partial co-integration on its own can no longer offer practical benefits to large players who need to invest in large companies like those forming part of the S&P 500.

6. Prediction of structural breaks

The previous section described the influence of the presence of random walk components in the spread that is used as basis for statistical arbitrage. Structural breaks occur when the co-integration relationship breaks down, causing the model error to be dominated by the random walk component and thus to fade away from the historical means. The likely result of a structural break is that a stop loss is reached, reducing the profitability of the strategy. It is therefore beneficial to be able to predict the probability of a structural break before a trade is entered. Lu et al. [9] developed a structural break-aware pairs trading strategy (SAPT) that includes two phases: phase 1 predicts the probability of structural breaks, while phase 2 optimize the strategy, taking the probability of structural breaks into consideration.

Estimating the probability of structural breaks is modeled as a binary classification problem. The objective function tries to minimize the cross-entropy of all co-integrated pairs over the trading period. To increase the amount of useful information based on which the probability of future structural breaks can be estimated, both time and frequency content of the spread is used. The Fourier transform is the most common method to extract frequency content from time information. As it is however based on the assumption of an infinitely long time window, it is not suited to extract frequency information that can be dynamically updated as the nature of the spread changes. Instead, a continuous wavelet transform is applied to the model error (or spread) to obtain both time and frequency information about historical spreads that continuously updates as the underlying price signals change. Lu et al. [9] used the Ricker wavelet to extract such 2D wavelet features that are then fed into a CNN. CNNs have been proven to be very efficient at extracting features from images. The CNN applies a set of convolution filters to the input data, followed by ReLU activation functions and by maximum pooling to prevent overfitting and limit computational effort.

In addition, Lu et al. [9] used an LSTM model that is applied directly to the spread and the price signals. LSTM is one of the most successful recurrent neural network structures to model time sequences, as it can model both short- and long-term memory. The LSTM produces time domain features that are then combined with the time-frequency features of the wavelet CNN. The concatenated time and frequency domain features are fed into a stack of fully connected layers; this allows possible nonlinear interactions between the features to be learned. The output of the last layer is a scalar that models the probability of a structural break. The model is trained by minimizing cross-entropy, using dropout to prevent overfitting.

Lu et al. [9] performed experiments, using stock tick data for the top 150 companies on the Taiwan Stock Exchange Capitalization Weighted Stock Index (TAIEX) to ensure liquidity. They trained the system on a daily basis, using the first 150 minutes of each day to extract co-integrated pairs and trading for the rest of the day till 5 minutes before closing time. 75% and 25% of the data are, respectively, used for training and testing, and the last 20% of the training data are for validation. Their evaluation metrics are the True, Missed, and False detection rates for breakpoints, as well as the average delay between the actual and detected breakpoints.

They compared the following alternative approaches to detect structural breaks:

- a statistical 3-std threshold,
- the ADF test for stationarity,
- Bayesian change-point detection,
- Using LSTM to predict the spread,

Performance measure	SWANet	Alternative methods
Missed detections	21%	27.4%
False detections	5.1%	34.6%
True detection rate	47.3%	24.4%
Average delay	17.3 min	22.3 min

Table 3.
Performance of SWANet compared to that of other structural breakdown detection methods.

- combining the use of the wavelet CNN in parallel with an LSTM network (SWANet).

They found that SWANet outperformed the alternative methods for structural breakdown detection. In **Table 3**, we provide the performance achieve by SWANet and the best of the alternative methods. The true detection rate can be further increased by increasing the tolerance for the accuracy by which a breakpoint must be detected—if the tolerance is increased from 20 to 70 minutes, the true detection rate increases from 47% to about 75%.

7. Reinforcement learning for strategy optimization

In the section on partial co-integration, we have shown that trading profits are sensitive to the trading strategy that is employed. This suggests the use of a formal optimization approach to maximize profits. Lu et al. [9] developed a strategy to optimize the levels of the spread to open and close trades and the trading volume for trades. They modeled the strategy as a Markov Decision Process (MDP) and proposed a deep reinforcement learning model to optimize the strategy by deciding the optimal boundaries. Their deep Q-network builds a Q-function based on historical events and estimates the Q-values by incorporating risks (e.g., the probability of structural breaks) as well as transaction costs.

The inputs fed into the Q-network include the following:

- The current spread;
- Current positions in the respective instruments;
- The last actions, where the set of possible actions is defined as all possible sets of trading boundaries that may be used;
- The probability of a structural break;
- The risk that the market will close (e.g., at end of day), forcing the trade to be closed prematurely;
- The reward associated with each possible action, taking into consideration change in normalized prices, volume, and transaction cost.

The outputs generated by the Q-network include the entry level, stop-loss closing level, and normal exit level. The Q-network has the objective of maximizing the Q-value, defined as the sum of expected rewards:

$$Q^*(s_t, a_t) = E_{s_{t+1}}\left[R(s_t, a_t, s_{t+1}) + \gamma \max_{a_{t+1}} Q(s_{t+1}, a_{t+1}) | s_t, a_t\right] \tag{25}$$

where R is the reward, Q^* is the maximum value of Q, and $\gamma \in [0, 1]$ is a factor discounting the maximum possible Q-values in future. The Q-network uses a stack of several fully connected layers to learn behavior. To achieve the objective of approximating the target Q-value, the following loss function is defined:

$$\mathcal{L}(\theta) = \left(R(s_t, a_t, s_{t+1}) + \gamma \max_{a_{t+1}} Q(s_{t+1}, a_{t+1}) | s_t, a_t - Q(s_t, a_t) \right)^2 \tag{26}$$

where θ represents all the trainable parameters in the Q-network. These parameters are updated after each training iteration using the gradient of the loss function:

$$\theta \leftarrow \theta - \eta \mathcal{L}(\theta) \tag{27}$$

thus updating the Q-values as follows:

$$Q(s_t, a_t) \leftarrow Q(s_t, a_t) + \eta \left[R(s_t, a_t, s_{t+1}) + \gamma \max_{a_{t+1}} Q(s_{t+1}, a_{t+1}) | s_t, a_t - Q(s_t, a_t) \right] \tag{28}$$

where η is the learning rate. The training is terminated when the loss function has reached a predetermined minimum value or when a predetermined maximum number of iterations is reached. The output is the action $a_t \in A$ that maximizes the Q-value, with A the set of all possible actions. Comparing Eq. (26) with Eq. (15), a similarity is observed between the operation of the Q-network and the way in which the Kalman filter is used to update the system states, with the Q-values playing the same role as system states and the difference between the current and maximum Q-value playing a similar role as new observations.

The performance of this strategy was measured on the same set of Taiwanese stocks as for the breakpoint detection discussed in the previous section. The evaluation metrics included cumulative net profit, maximum drawdown, and Sharpe ratio. They compared a Q-network that uses only trading and stop-loss boundaries with a network that also factors in the probability of structural breaks, market closing risk, and transaction costs. The latter approach outperformed the former approach by about 15% over a trading period of 2.5 years, displaying the benefit of anticipating the occurrence of a structural break that will cause the spread to revert from mean reversion to a random walk. Predicting structural breaks had an even bigger impact on the reduction of risk, lowering maximum drawdown from 16.9% to 2%, and increasing the Sharpe ratio from 1.01 to 4.30 and the Sortino ratio from 1.41 to 13.18.

8. Conclusions

This chapter illustrated the applicability of the Kalman filter approach to the optimization of financial investment decisions. Financial markets represent a relevant example of a complex system of which the states can only be partly observed. As the quoted prices of instruments on an exchange are influenced by many factors, and as not all market players have access to the same information or may be restricted in terms of their speed of their reactions to new information, the true value of such instruments will not always correspond to the spot price. Statistical arbitrage intends to exploit temporary mispricings observed in the market by anticipating how the market will react to restore prices toward true values. The concept of co-integration was developed to provide a formal basis for pairs trading, which may be regarded as the initial form of statistical arbitrage. As the spreads that are created from sets of co-integrated instruments cannot be guaranteed to be mean reverting, co-integration had to be extended to provide for both mean-reverting and random walk system states, resulting in the concept of partial co-integration. This left the challenge to determine how the spread is divided between the mean-reverting and random walk components. The Kalman filter algorithm proved to be suitable to solve for these unknown system states, allowing the accuracy and profitability of trading strategies based on statistical arbitrage to be improved. Recent research has provided evidence that the Kalman filter method applied to partial co-integration can be complemented by using state-of-the-art AI methods, including convolutional neural networks and reinforcement learning. CNN classifiers help to address the weaknesses of partial co-integration by predicting when the random walk system state is likely to start dominating, thus avoiding possible loss-making trades. Reinforcement learning provides a systematic approach to make optimal selections for trading parameters, including the levels at which trades should be entered and exited. This provides a good example of how long-established methods can be combined with more recent innovations to meet the challenges of an ever-increasing complex world.

Author details

Alwyn J. Hoffman
School of Electrical, Electronic and Computer Engineering, North-West University, Potchefstroom, South Africa

*Address all correspondence to: alwyn.hoffman@nwu.ac.za

References

[1] Kalman RE. A new approach to linear filtering and prediction problems. Transactions of the ASME -Journal of Basic Engineering. 1960;**82**((Series D)): 35-45

[2] Markowitz H. Portfolio selection. Journal of Finance. 1952;7(5):77-91

[3] Ross SA. The arbitrage theory of capital asset pricing. Journal of Economics. 1976;**13**(3):341-360

[4] Lo AW. Hedgefunds: An Analytic Perspective. Princeton: Princeton University Press; 2010. Available from: http://site.ebrary.com/id/10394786

[5] Vidyamurthy G. Pairs Trading : Quantitative Methods and Analysis. Hoboken, NJ: John Wiley; 2004

[6] Johansen S. Likelihood-Based Inference in Cointegrated Vector Autoregressive Models. Oxford: Clarendon Press; 1995

[7] Clegg M, Krauss C. Pairs trading with partial cointegration. Quantitative Finance. 2018;**18**(1):121-138. DOI: 10.1080/14697688.2017.1370122

[8] Hoffman A. Statistical arbitrage on the JSE based on partial co-integration. Investment Analysts Journal. 2021;**50**(2): 110-132. DOI: 10.1080/10293523.2021.1886723

[9] Lu J-Y, Lai H-C, Shih W-Y, Chen Y-F, Huang S-H, Chang H-H, et al. Structural break-aware pairs trading strategy using deep reinforcement learning. The Journal of Supercomputing. 2022;**78**(3): 3843-3882. DOI: 10.1007/s11227-021-04013-x

[10] Avellaneda M, Lee J-H. Statistical arbitrage in the US equities market. Quantitative Finance. 2010;**10**(7):761-782. DOI: 10.1080/14697680903124632

[11] Thorp E. A Mathematician on Wall Street. 2000. Available from: http://www.wilmott.com/pdfs/080617_thorp.pdf

[12] Bhave A, Libertini N. A Study of Short Term Mean Reversion in Equities. 2013. Available from: http://www.361capital.com/wp-content/uploads/2013/10/361_Capital_Study_Short_Term_Mean_Reversion_Equities.pdf

[13] Engle RF, Granger CWJ. Co-integration and error correction: Representation, estimation and testing. Econometrica. 1987;**55**(2):251-276

[14] Sjo B. Testing for Unit Roots and Cointegration. 2008. Available from: https://www.iei.liu.se/nek/ekonometrisk-teori-7-5-hp-730a07/labbar/1.233753/dfdistab7b.pdf

[15] Phillips P, Ouliaris S. Asymptotic properties of residual based tests for. Econometrica: Journal of the Econometric Society. 1990;**58**(1):165. DOI: 10.2307/2938339

[16] Gatev E, Goetzmann WN, Rouwenhorst KG. Pairs trading: Performance of a relative-value arbitrage rule. The Review of Financial Studies. 2006;**19**(3):797-827. DOI: 10.1093/rfs/hhj020

[17] Thorp E. A Perspective on Quantitative Finance: Models for Beating the Market. 2003. Available from: http://media.wiley.com/product_data/excerpt/11/04700235/0470023511.pdf

[18] Zhou ZZ, Kang S. Impact of the Shanghai-Hongkong stock connect on

the performance of A-H pairs trading. In: Proceedings of the International Conference on Information Systems (ICIS). 2019. pp. 1-16

[19] Vidyamurthy G. Pairs Trading: Quantitative Methods and Analysis. Hoboken, New Jersey: J. Wiley; 2004

[20] Rudy J. Four Essays in Statistical Arbitrage in Equity Markets. Liverpool, U.K.: John Moores University; 2011

[21] Chan EP. Algorithmic Trading : Winning Strategies and their Rationale. Hoboken, New Jersey: John Wiley & Sons, Inc.; 2013

[22] Fuller WA. Introduction to Statistical Time Series. 2nd ed. New York: Wiley; 1996

[23] Harlacher M. Cointegration Based Statistical Arbitrage. [Online]. 2012. Available from: https://stat.ethz.ch/research/mas_theses/2012/harlacher.pdf [Accessed: July 14, 2015]

[24] Clegg M. Modeling time series with both permanent and transient components using the partially autoregressive model. 2015. SSRN 2556957.10.2139/ssrn.2556957

[25] Mashele H, Terblanche S, Venter J. Pairs trading on the Johannesburg stock exchange. Investment Analysts Journal. 2013;**2013**(78):13-26

[26] Sharpe W, Alexander G. Investments. Englewood: Prentice Hall; 1990

Section 3

Applications in Dynamic and Stochastic Systems

Chapter 7

Perspective Chapter: Insights from Kalman Filtering with Correlated Noises Recursive Least-Square Algorithm for State and Parameter Estimation

Abd El Mageed Hag Elamin Khalid

Abstract

This article explores the estimation of parameters and states for linear stochastic systems with deterministic control inputs. It introduces a novel Kalman filtering approach called *Kalman Filtering with Correlated Noises Recursive Generalized Extended Least Squares (KF-CN-RGELS) algorithm*, which leverages the cross-correlation between process noise and measurement noise in Kalman filtering cycles to jointly estimate both parameters and system states. The study also investigates the theoretical implications of the correlation coefficient on estimation accuracy through performance analysis involving various correlation coefficients between process and measurement noises. The research establishes a clear relationship: the accuracy of identified parameters and states is directly proportional to positive correlation coefficients. To validate the efficacy of this algorithm, a comprehensive comparison is conducted among different algorithms, including the standard Kalman filter algorithm and the augmented-state Kalman filter with correlated noises algorithm. Theoretical findings are not only presented but also exemplified through a numerical case study to provide valuable insights into practical implications. This work contributes to enhancing estimation accuracy in linear stochastic systems with deterministic control inputs, offering valuable insights for control system design and state-space modeling.

Keywords: correlated noises, least squares, linear stochastic system, parameter estimation, estimation accuracy

1. Introduction

Finding a state-space model is crucial for designing control systems. A big part of this process involves changing the unknown parameters in a controller's transfer function or state-space representation to get the desired stability, performance, and robustness [1]. In the field of control, system identification has been done using many different estimation methods, such as least-squares methods [2], iterative identification methods [3], Bayesian methods [4, 5], separated least-squares methods [6, 7], and maximum likelihood methods [8]. However, identifying the parameters of state-space models is more difficult. For instance, a state-space model has unknown states as well as unknown parameter matrices or vectors [9, 10]. In this area, some studies focus on identifying nonlinear state-space models [11]. Meanwhile, other studies look at how to identify linear state-space models. For example, Safarinejadian et al. [12] introduced a new way to identify a state-space model for a single input single-output fractional-order system based on a new fractional-order Kalman filter with correlated noises. To find the parameters of linear state-space models, Yu et al. [13] used the least-squares estimation framework and made the Hankel matrix factorization less affected by Markov-parameter estimation error by using a single optimization framework instead of the two-step method used by Yu et al. [14] to identify the structured system matrices. A study [8] introduced two hidden variables and used the expectation-maximization algorithm to estimate time delays and parameters in a state-space model with unknown time delay. To identify slowly changing linear time-varying systems, Razmjooei et al. [15] used a two-step method: first, they estimated parameters using Legendre basis functions, then used the Kalman filter with those estimated parameters to compute the system's states. Industrial processes often suffer from noise in both measurements and the processes themselves, making accurate state and parameter estimation challenging. To address this, Li et al. [16] applied data filtering to mitigate the impact of colored noise in the measurement equation for bilinear state-space models. Recognizing the influence of both process and measurement noise, Cui et al. [17] leverages data filtering to mitigate the effects of colored noise, specifically within the state equations. Aiming to reduce the impact of colored noise and achieve better estimation accuracy, Wang et al. [18] proposed a novel algorithm. This approach combines filtering techniques with a recursive generalized least-squares method, utilizing filtered measurement data for continuous updates [19]. Ma et al. [20] focused on identifying state-space models in a specific form (observer canonical) while dealing with white noise affecting the output measurements. Building upon existing methods, Cui et al. [21] presented a novel algorithm for jointly estimating parameters and states in a specific class of state-space models (observer canonical) that are exposed to colored noise. This approach leverages both Kalman filtering and gradient search techniques for improved accuracy. In Cui et al. [22], researchers tackle a specific noise situation (white noise in state equations, moving average noise in measurements) by developing two algorithms for jointly estimating states and parameters in state-space systems. These algorithms rely on the "auxiliary model" concept. Motivated by the limitations of existing methods, this study investigates state-space models experiencing specific noise characteristics (white noise, moving average noise) and correlated noise behaviors, aiming to develop robust estimation algorithms for such systems. In the field of parameter and state estimation, researchers are striving to achieve both more accurate results and faster calculations. Multi-innovation identification theory tackles this challenge by utilizing more system and data information to improve parameter estimation accuracy, as shown in previous studies [23–28]. Building upon past studies of correlated noise, this work takes a deeper dive into the specific effects of varying correlation coefficients on estimation accuracy. This knowledge will inform the development of improved estimation algorithms for state-space systems under realistic noise conditions.

This paper makes four main contributions:

1. The paper challenges the conventional assumption of the Kalman filter, which relies on known system parameters. It uses past estimates of the system's

parameters to predict its current state and then uses those predictions to improve the estimates of those parameters.

2. In the context of correlated process noise and measurement noise, the paper introduces a new formulation of the Kalman filter algorithm that preserves the fundamental assumption of Kalman filtering, which operates with uncorrelated noise.

3. Emphasizing the significant role of the correlation coefficient between process noise and measurement noise, the paper demonstrates that higher correlation coefficients lead to more accurate estimates.

4. The negative correlation coefficient reduces the estimation and identification accuracy, considering accuracy from two different perspectives: observation and model. It leads to an increase in measurement covariance and cross-covariance between state and process noise. This increase in covariance and cross-covariance affects the filtering results, leading to less accurate state estimates and parameter estimation.

This article is structured as follows. Section 2 describes the system model of a linear stochastic state-space system. Section 3 contains the formulation of the Kalman filter for handling cross-correlated noise. Section 4 presents the algorithm used for the comparative evaluation. Section 5 demonstrates the formulation of the identification model for linear stochastic state-space models. Section 6 contains the derivation of the proposed algorithm (KF-CN-RGELS). Section 7 explains the impact of negative correlation coefficients. Section 8 provides an example to verify the effectiveness of the proposed algorithm. Concluding observations can be found in Section 9.

2. The system model for linear stochastic system

Let us introduce some notation. The expressions "F=:X" or "X:=F" indicate that "F" is defined as "X." The symbol "q" represents a unit back-shift operator, where $q^{-1}x(t)$ denotes $x(t-1)$. The superscript "T" denotes the transpose of vectors/matrices [29].

Consider the following linear-stochastic system:

$$x(t+1) = Fx(t) + Gu(t) + w(t), \tag{1}$$

$$y(t) = Hx(t) + du(t) + v^*(t), \tag{2}$$

$$v^*(t) := \left(1 + J_1 q^{-1} + J_2 q^{-2} + \cdots + J_{n_J} q^{-n_J}\right) v(t), \tag{3}$$

Here, $u(t) \in \mathbb{R}$ and $y(t) \in \mathbb{R}$ are the input and output of the system, respectively. The state vector of the system is denoted as $x(t) := [x_1(t), \ldots, x_n(t)]^T \in \mathbb{R}^n$. The white noise process $v(t) \in \mathbb{R}$ has zero mean and variance σ_v^2, and $w(t) := [w_1(t), \ldots, w_n(t)]^T \in \mathbb{R}^n$ represents the white noise vector with zero mean. $J(q)$ in the unit back-shift operator q^{-1} is defined as $J(q) = 1 + J_1 q^{-1} + J_2 q^{-2} + \cdots + J_{n_J} q^{-n_J} \in \mathbb{R}$. The matrices F, G, and H representing system parameters are defined as follows:

$$F := \begin{bmatrix} -f_1 & 1 & 0 & \cdots & 0 \\ -f_2 & 0 & 1 & \cdots & 0 \\ \vdots & \vdots & \vdots & \ddots & \vdots \\ -f_{n-1} & 0 & 0 & \cdots & 1 \\ -f_n & 0 & 0 & \cdots & 0 \end{bmatrix} \in \mathbb{R}^{n\times n}, G := \begin{bmatrix} g_1 \\ g_2 \\ \vdots \\ g_{n-1} \\ g_n \end{bmatrix} \in \mathbb{R}^n, H := [1, \quad 0, \quad \cdots \quad 0] \in \mathbb{R}^{1\times n} \quad (4)$$

3. Formulation of Kalman filter with noises cross-correlation

In this section, we delve into the formulation of a Kalman filter designed to address the presence of noise cross-correlation in linear stochastic state-space models. We make several key assumptions and present the mathematical derivations essential for a clear understanding of this formulation.

Assumptions: Consider a linear stochastic state-space model described by Eqs. (1) and (2). In this framework:

- The sequences $w(t)$ and $v(t)$ are zero-mean Gaussian white noise processes.
- The variance of $w(t)$ at time t is represented by $Q(t)$, which is a positive definite matrix.
- The variance of $v(t)$ at time t is denoted as $R(t)$, also a positive definite matrix.
- These noise sequences, $w(t)$ and $v(t)$, exhibit statistical correlation. Specifically, the correlation is defined by equation [30, 31]:

$$E\left[w(k)v(l)^T\right] = S(k)\delta_{kl}, , k, l = 0, 1, \ldots \quad (5)$$

where δ_{kl} represents the Kronecker delta function, and each $S(k)$ is a non-negative definite matrix.

Formulation: To effectively deal with correlated noise sequences, a gain matrix T is introduced and incorporated into the system equations. This process involves adding the gain matrix T to the measurement Eq. (2). The modified system Eq. (1) now reads:

$$x(t+1) \quad = Fx(t) + Gu(t) + w(t) + T[y(t) - Hx(t) - du(t) - J(q)v(t)] \quad (6)$$

To simplify the representation, a noise term $v^*(t)$ is defined as follows:

$$v^*(t) := \left(1 + J_1 q^{-1} + J_2 q^{-2} + \cdots + J_{n_J} q^{-n_J}\right) v(t) \quad (7)$$

As a result, the modified system equation becomes:

$$x(t+1) \quad = (F - TH)x(t) + (G - Td)u(t) + Ty(t) + [w(t) - Tv^*(t)]. \quad (8)$$

In Eq. (8), the term $Ty(t)$ represents a determined control input. To further simplify, we introduce the following notations:

$$\overline{F}:=(F - TH), \tag{9}$$

$$\overline{G}:=(G - Td), \tag{10}$$

$$\overline{w}(t):=w(t) - Tv^*(t) \tag{11}$$

With these notations, the modified system dynamics equation can be succinctly rewritten as:

$$x(t+1) = \overline{F}x(t) + \overline{G}u(t) + Ty(t) + \overline{w}(t). \tag{12}$$

3.1 Derivation of the gain matrix T

From Eq. (11), the mean and covariance of $\overline{w}(t)$ can be described as follows:

$$E[\overline{w}(t)] = E[w(t)] - TE[v^*(t)] = 0, \tag{13}$$

$$E\left[\overline{w}(t)\overline{w}^T(t)\right] = \overline{Q}. \tag{14}$$

Notably, we have assumed that the noises are correlated at the same time. By employing Eq. (7), we can demonstrate the relationship:

$$E\left[w(t)v^T(t)\right] = E\left[w(t)v^{*T}(t)\right] = S. \tag{15}$$

Substituting Eq. (11) into (14), we obtain the following expression for $\overline{Q}$:

$$\overline{Q} = Q + (TR - S)T^T - TS^T. \tag{16}$$

Taking the transpose of Eq. (11), we define:

$$\overline{w}^T(t) = w^T(t) - v^{*T}(t)T^T \tag{17}$$

When we multiply Eq. (11) by $\overline{w}^T(t)$ and take the expectation, we get:

$$E\left[\overline{w}(t)\overline{w}^T(t)\right] = E\left[\overline{w}(t)w^T(t)\right] - E\left[\overline{w}(t)v^{*T}(t)\right]T^T \tag{18}$$

In line with the standard Kalman filter assumptions, we assume that the new process noise, $\overline{w}(t)$, is independent of the measurement noise $v^*(t)$, meaning $E\left[\overline{w}(t)v^{*T}(t)\right] = 0$. With this assumption, we arrive at the relationship:

$$E\left[\overline{w}(t)\overline{w}^T(t)\right] = E\left[\overline{w}(t)w^T(t)\right] = \overline{Q} \tag{19}$$

By right-multiplying Eq. (11) by $w^T(t)$ and taking the expected value, we can establish:

$$E\left[\overline{w}(t)w^T(t)\right] = E\left[w(t)w^T(t) - TE\left[w^T(t)v^*(t)\right]\right. \tag{20}$$

This relationship leads to the equation:

$$\overline{Q} = Q - TS^T \tag{21}$$

By subtracting Eq. (21) from (16), we can derive the following relationship:

$$TR - S = 0, \text{ since } T \neq 0 \tag{22}$$

Hence, the gain matrix T needed to handle the uncorrelated process noise $\overline{w}(t)$ and the measurement noise $v^*(t)$ can be calculated as:

$$T = SR^{-1} \tag{23}$$

Remark 1: It is crucial to consider the new variance $\overline{Q} = Q - TS^T$ of the noise sequence $\overline{w}(t)$. This is a critical adjustment to the model, as it departs from the variance Q of the noise sequence $w(t)$.

3.2 Kalman filter prediction and correction cycles

At this point, we focus on the system dynamics equation and the measurement equation, which are given in Eqs. (24)–(25):

$$x(t+1) = \overline{F}x(t) + \overline{G}\,u(t) + Ty(t) + \overline{w}(t) \tag{24}$$

$$y(t) = Hx(t) + du(t) + v^*(t) \tag{25}$$

The prediction and correction cycles of the modified Kalman filter for this proposed system are as follows:

Prediction cycle

For $x_p(t)$ representing the prediction state, the prediction cycle equations are:

$$x_p(t+1) = \overline{F}x_p(t) + \overline{G}\,u(t) + Ty(t) \tag{26}$$

$$P_p = \overline{F}P_p\overline{F}^T + \overline{Q} \tag{27}$$

$$K = P_pH^T\left(HP_pH^T + R\right)^{-1} \tag{28}$$

Here, K and P_p represent the Kalman gain and the estimation error covariance, respectively.

Correction cycle

For $x_c(t)$ be the corrected state or the estimated state, the correction cycle equations will be as follows.

$$x_c(t+1) = x_p(t+1) + Kv^*(t) \tag{29}$$

(where) $v^*(t) = y(t) - c * x(t) - d * u(t)$

The estimated error covariance is calculated as:

$$P_c = (I_n - KH)P_p, \tag{30}$$

Eqs. (26)–(30) describe the prediction and correction cycles of the Kalman filtering process when dealing with correlated noise sequences.

4. Introduction and formulation of Kalman filtering algorithms for comparative assessment

In this section, we introduce and formulate two Kalman filtering algorithms for the purpose of conducting a comprehensive comparative analysis with the proposed algorithm. The algorithms to be compared are the standard Kalman filter (SKF) and the State Augmented Kalman Filter with Correlated Noises (AUG-KF). Each of these algorithms is presented with its assumptions and key equations for prediction and correction cycles [32–34].

4.1 Standard Kalman filter (SKF)

Consider the system model described by Eqs. (1)–(2)

Assumptions: The process noise ($w(t)$) and measurement noise ($v(t)$) are independent.

Formulation: Prediction and Correction Cycles

Prediction cycle

Predicted state: $x_p(t+1) = Fx_p(t) + G\,u(t)$

Prediction error covariance: $P_p = FP_pF^T + Q$

Kalman gain: $K = \left(P_pH^T\right)\left(HP_pH^T + R\right)^{-1}$

Correction cycle

Corrected state: $x_c(t+1) = x_p(t+1) + Kv^*(t)$, Where $v^*(t) = y(t) - c * x(t) - d * u(t)$.

Correction error covariance: $P_c = P_p - K\left(P_pH^T\right)^T$

4.2 Augmented-State Kalman filter (AUG-KF)

State augmentation is a technique to extend the state vector of a system by adding some auxiliary variables that are related to the original state or the noise. In [35], the authors augment the state vector by appending the process noise vector, so that the cross-correlated noises can be treated as independent noises in the augmented system. This technique can simplify the filter design and improve the estimation accuracy, but it also increases the dimension of the state vector and the computational cost of the filter. With reference to method used in [35] we can derive the AUG-KF algorithm for the system (1)–(2) as

Assumptions: The process noise ($w(t)$) and measurement noise ($v(t)$) are correlated with $E\left[w(t)v^*(t)^T\right] = S$ represents the cross-covariance between process noise $w(t)$ and measurement noise $v^*(t)$.

Formulation: Initialization, Prediction and Update Cycles:

Initialization:

$n_t = y(t) - H\,x(t) - d\,u(t) - J_1\,v(t-1) - J_2\,v(t-2)$, where n_t is the actual observation.

Augmented Kalman Gains:

$k_a = [k_x, k_w]$, where $k_x = P\,H^T\left(HP_pH^T + R\right)^{-1}$, $k_w = S\,\left(HP_pH^T + R\right)^{-1}$.

Update process noise and state estimate:

$w(\,t) = k_w\;n_t$, $x_p(t+1) = Fx_p(t) + G\,u(t) + w(\,t)$.

$$P_p(\mathrm{t}+1) = F\,P_p(\mathrm{t})F^T + P_w + F\,P_{xw} + P_{wx}\,F^T$$

Update covariance matrix of augmented state and process noise:

$P_{Aug} = \begin{bmatrix} P_c(\mathrm{t}) & P_{xw} \\ P_{wx} & P_w \end{bmatrix}$, where $P_c(\mathrm{t}) = P_p(\mathrm{t}+1) - k_x\left(HP_p(t+1)H^T + R\right)^{-1} {k_x}^T$,

$P_w = Q - k_w\left(HP_c(t)H^T + R\right)^{-1} {k_w}^T$, $P_{wx} = -k_w\left(HP_c(t)H^T + R\right)^{-1} {k_x}^T$

And $P_{xw} = -k_x\left(HP_c(t)H^T + R\right)^{-1} {k_w}^T$.

Update for the next time step:

Corrected state:

$$x_c(\,t+1) = x_p(t+1) + k_x\;n_t,$$

Correction error covariance:

$$P_c(t+1) = P_c(t) - k_x\left(HP_c(t)H^T + R\right)^{-1} {k_x}^T.$$

These algorithms will form the foundation for a comparative analysis in the following sections. We will assess their performance and effectiveness in managing noise correlations, comparing them to the proposed algorithm. This evaluation aims to identify the most effective approach in various scenarios.

5. The identification model for linear stochastic system

Let us define some essential notation:

- "I_n" denotes an identity matrix of appropriate size, typically $n \times n$.
- "$\mathbf{1}_n$" represents an n-dimensional column vector with all elements equal to unity.
- "$\hat{\theta}(t)$ "represents the estimate of θ at time t, and "$\hat{x}(t)$" denotes the estimate of $x(t)$.

Now, based on Section 3, we modify the system equations:

$$x(t+1) = \overline{F}x(t) + \overline{G}u(t) + Ty(t) + \overline{w}(t) \tag{31}$$

$$y(t) = Hx(t) + du(t) + v^*(t) \tag{32}$$

With

$$v^*(t) := \left(1 + J_1q^{-1} + J_2q^{-2} + \cdots + J_{n_J}q^{-n_J}\right)v(t)$$

$\overline{F} := (F - TH)$, $\overline{w}(t) := w(t) - Tv^*(t)$.

The system's input and output are represented by $u(t) \in R$ and $y(t) \in R$, respectively, with $x(t) := [x_1(t), \ldots, x_n(t)]^T \in \mathbb{R}^n$ represents the system state vector. $v(t) \in R$ is a white noise process with zero mean and variance σ_v^2, and $w(t) := [w_1(t), \ldots, w_n(t)]^T \in \mathbb{R}^n$

denotes the white noise vector with zero mean. The polynomial $J(q)$ in the unit backshift operator q^{-1} is expressed as $J(q) = 1 + J_1 q^{-1} + J_2 q^{-2} + \cdots + J_{n_J} q^{-n_J} \in \mathbb{R}$. We also have system parameters: $F \in \mathbb{R}^{n\times n}, G \in \mathbb{R}^n, H \in \mathbb{R}^{1\times n}$ and $d \in (\mathbb{R})$

The matrices F, G and H are defined as follows:

$$\overline{F} := \begin{bmatrix} -\overline{f}_1 & 1 & 0 & \cdots & 0 \\ -\overline{f}_2 & 0 & 1 & \cdots & 0 \\ \vdots & \vdots & \vdots & \ddots & \vdots \\ -\overline{f}_{n-1} & 0 & 0 & \cdots & 1 \\ -\overline{f}_n & 0 & 0 & \cdots & 0 \end{bmatrix} \in \mathbb{R}^{n\times n},$$
$$\overline{G} := \begin{bmatrix} \overline{g}_1 \\ \overline{g}_2 \\ \vdots \\ \overline{g}_{n-1} \\ \overline{g}_n \end{bmatrix} \in \mathbb{R}^n, H := [1, \quad 0, \quad \cdots \quad 0] \in \mathbb{R}^{1\times n} \tag{33}$$

Assuming that $y(t), u(t), w(t)$, and $v(t)$ are strictly proper, meaning their values are zero for 0 for $t \leq 0$, and that the orders n and n_J are known, we can derive the following state equations from (1)–(3):

$$\begin{aligned} x_1(t) &= -\overline{f}_1 x_1(t-1) + x_2(t-1) + \overline{g}_1 u(t-1) \\ &\quad + \overline{w_1}(t-1) + T_1 y(t-1) \\ x_2(t-1) &= -\overline{f}_2 x_1(t-2) + x_3(t-2) + \overline{g}_2 u(t-2) \\ &\quad + \overline{w_2}(t-2) + T_2 y(t-2) \\ x_3(t-2) &= -\overline{f}_3 x_1(t-3) + x_4(t-3) + \overline{g}_3 u(t-3) \\ &\quad + \overline{w_3}(t-3) + T_3 y(t-3) \\ &\vdots \\ x_{n-1}(t-n+2) &= -\overline{f}_{n-1} x_1(t-n+1) + x_n(t-n+1) \\ &\quad + \overline{g}_{n-1} u(t-n+1) + \overline{w}_{n-1}(t-n+1) + T_{n-1} y(t-n+1) \\ x_n(t+1-n) &= -\overline{f}_n x_1(t-n) + \overline{g}_n u(t-n) + \overline{w}_n(t-n) + T_n y(t-n). \end{aligned} \tag{34}$$

From these n equations, we can express $x_1(t)$ as:

$$x_1(t) = -\sum_{i=1}^{n} \overline{f}_i x_1(t-i) + \sum_{i=1}^{n} \overline{g}_i u(t-i) + \sum_{i=1}^{n} \overline{w}_i(t-i) + \sum_{i=1}^{n} T_i y(t-i) \tag{35}$$

Substituting $H = [1, 0, \ldots, 0]$ in (2), we obtain:

$$y(t) = x_1(t) + du(t) + \left[1 + J_1 q^{-1} + \cdots + J_{n_J} q^{-n_J}\right] v(t) \tag{36}$$

Now, let us define the parameter vector θ and the information vector $\varphi(t)$:

$$\overline{\theta}:=\left[\overline{F}^T, \overline{G}^T, d, J^T\right]^T \in \mathbb{R}^{n_s}, n_s = 2n + 1 + n_J, \tag{37}$$

$$\overline{F}:=\left[\overline{f}_1, \overline{f}_2, \dots, \overline{f}_n\right]^T \in \mathbb{R}^n, \tag{38}$$

$$J:=\left[J_1, J_2, \dots, J_{n_J}\right]^T \in \mathbb{R}^{n_J}, \tag{39}$$

$$\varphi(t):=\left[\varnothing_{\overline{f}}(t)^T, \varnothing_{\overline{g}}(t)^T, u(t), \varnothing_v(t)^T\right]^T \in \mathbb{R}^{n_s} \tag{40}$$

Where:

$$\varnothing_{\overline{f}}(t):=[-x_1(t-1), \dots, -x_1(t-n)]^T \in \mathbb{R}^n$$

$$\varnothing_{\overline{g}}(t):=[u(t-1), u(t-2), \dots, u(t-n)]^T \in \mathbb{R}^n$$

$$\varnothing_v(t):=\left[v(t-1), v(t-2), \dots, v\left(t-n_J\right)\right]^T \in \mathbb{R}^{n_J}$$

Now, we introduce the variables $\gamma(t)$, and $\beta(t)$ defined as:

$$\gamma(t):=\sum_{i=1}^{n} w_i(t-i) = w_1(t-1) + w_2(t-2) + \dots + w_n(t-n) \tag{41}$$

$$\beta(t):=\sum_{i=1}^{n} T_i y(t-i) = T_1 y(t-1) + T_2 y(t-2) + \dots + T_n y(t-n) \tag{42}$$

Using (35) and (41), Eq. (36) can be rewritten as:

$$\begin{aligned} y(t) &= \varnothing_{\overline{f}}(t)^T \overline{F} + \varnothing_{\overline{g}}(t)^T \overline{G} + \gamma(t) + \beta(t) + u(t) + \varnothing_v(t)^T J + v(t) \\ &= \varphi(t)^T \theta + \gamma(t) + \beta(t) + v(t) \end{aligned} \tag{43}$$

Eq. (43) represents the identification model for a linear stochastic state-space system as defined in (1)–(2). The main objective of this paper is to present an algorithm that jointly estimates the system states and unknown parameters using recursive generalized extended least squares. Additionally, we aim to investigate the impact of correlations between process and measurement noises on estimation accuracy.

Remark 2: To simplify the identification process, the observable general state-space system described in (1)–(2) is transformed into the observer canonical form. This transformation serves to reduce the number of parameters that need to be identified, making the estimation process more efficient and accurate.

6. The KF-CN-RGELS algorithm

This section introduces the KF-CN-RGELS algorithm for the joint estimation of system parameters and states in a canonical observer state-space system (Eqs. 1 and 2).

The algorithm comprises two key components: the parameter estimation algorithm and the state estimation algorithm. These algorithms are developed to address the challenge of estimating parameters and states in the proposed system, and when combined, they provide a comprehensive solution.

6.1 The parameter estimation algorithm

The parameter estimation algorithm is centered around minimizing the quadratic criterion function defined as:

$$C(\theta):=\sum_{j=1}^{L}\|y(j)-\varphi(j)^T\theta-\gamma(j)-\beta(j)\|^2, \tag{44}$$

This minimization process helps estimate the system parameters based on the identification model (Eq. 43) using the least-squares principle [22]. The parameter estimation is performed recursively, and it can be expressed as follows:

$$\hat{\theta}(t)=\hat{\theta}(t-1)+L(t)\left[y(t)-\gamma(t)-\beta(t)-\varphi(t)^T\hat{\theta}(t-1)\right] \tag{45}$$

$$L(t)=\frac{P(t-1)\varphi(t)}{1+\varphi(t)^T P(t-1)\varphi(t)} \tag{46}$$

$$P(t)=P(t-1)-L(t)[P(t-1)\varphi(t)]^T, P(0)=p_0 I_n \tag{47}$$

However, in implementing these algorithms, certain challenges arise due to unknown process and measurement noise sequences, as well as unmeasurable states. To address these challenges, the concept of auxiliary models is introduced to replace unknown parameters and states with their estimates. In this modified parameter estimation algorithm, $\hat{\varphi}(t)$ and $\hat{\gamma}(t)$ are used instead of $\varphi(t)$ and $\gamma(t)$, resulting in the following algorithm [22]:

$$\hat{\theta}(t)=\hat{\theta}(t-1)+L(t)\left[y(t)-\hat{\gamma}(t)-\beta(t)-\hat{\varphi}(t)^T\hat{\theta}(t-1)\right] \tag{48}$$

$$L(t)=\frac{P(t-1)\hat{\varphi}(t)}{1+\hat{\varphi}(t)^T P(t-1)\hat{\varphi}(t)} \tag{49}$$

$$P(t)=P(t-1)-L(t)[P(t-1)\hat{\varphi}(t)]^T, P(0)=p_0 I_n \tag{50}$$

The vectors $\hat{\phi}_v(t)$, $\hat{\gamma}(t)$ and $\beta(t)$ are formed using the output, the process noise, and measurement noise sequences, which are calculated using the equations provided.

To form $\hat{\phi}_v(t)$ in $\hat{\varphi}(t)$ and $\hat{\gamma}(t), \hat{w}(t-i)$ and, $\hat{v}(t-i)$ can be calculated using (1) and (2) as

$$\hat{\overline{w}}(t)=\hat{x}(t+1)-\hat{\overline{F}}(t)\hat{x}(t)-\hat{\overline{G}}(t)u(t)-T\left(cx(t)+\hat{d}\,u(t)\right) \tag{51}$$

$$\widehat{v^*}(t)=y(t)-H\hat{x}(t)-\hat{d}(t)u(t), \tag{52}$$

$$\hat{v}(t)=\widehat{v^*}(t)-\hat{J}_1\hat{v}(t-1)-\hat{J}_2\hat{v}(t-2) \tag{53}$$

6.2 The state estimation algorithm

This section deals with the problem of estimating non-measurable states of the information vector $\varphi(t)$. To overcome this challenge, a modified Kalman filter prediction and update cycle, described in Section 4, is used to estimate the system state. The state estimation depends on the degree of correlation between process noise and measurement noise in the proposed system. The state estimation algorithm is as follows:

Prediction cycle:

$$xp(t+1) = \hat{\overline{F}}xp(t) + \hat{\overline{G}}u(t) + Ty(t); \tag{54}$$

$$P = \hat{\overline{F}}P\hat{\overline{F}}^T + Q - \hat{T}\hat{S}^T \tag{55}$$

Calculation of Kalman gain:

$$K = (Pc)/c\,Pc^T + \hat{R} \tag{56}$$

Correction cycle:

$$x(t+1) = xp(t+1) + +K\widehat{v^*}(t) \tag{57}$$

$$P = (eye(n) - Kc)P \tag{58}$$

In these equations, $\overline{F} = (F - TH), \overline{Q} = Q - TS^T$ and $S = \rho_{w,v}\sqrt{R}\sqrt{Q}$, as stated in Section 4. Therefore,

$$\hat{x}(t) = x_c(t) \tag{59}$$

where $x_p(t)$ and $x_c(t)$ are predicted and corrected states at time t, (respectively)

Remark 3: A Kalman filter estimates the state of a system, assuming that its parameters are known. To overcome this issue, the concept of an auxiliary model that replaces all system parameters with estimates in the prediction and correction cycles of the modified Kalman filter recursion equations, as illustrated in (55)–(60), is applied.

6.3 The joint parameter and state estimation algorithm

The joint parameter and state estimation algorithm combines the parameter estimation and state estimation algorithms to recursively estimate both the system's parameter vector and state vector. The algorithm is described by the following equations:

$$\hat{\theta}(t) = \hat{\theta}(t-1) + L(t)\left[y(t) - \hat{\gamma}(t) - \beta(t) - \hat{\varphi}(t)^T\hat{\theta}(t-1)\right] \tag{60}$$

$$L(t) = \frac{P(t-1)\hat{\varphi}(t)}{1 + \hat{\varphi}(t)^T P(t-1)\hat{\varphi}(t)} \tag{61}$$

$$P(t) = P(t-1) - L(t)[P(t-1)\hat{\varphi}(t)]^T, P(0) = p_0 I_n \tag{62}$$

$$\hat{\varphi}(t) = \left[\hat{\varnothing}_{\bar{f}}(t)^T, \varnothing_g(t)^T, u(t), \hat{\varnothing}_v(t)^T\right]^T \in \mathbb{R}^{n_s}, \tag{63}$$

$$\hat{\varnothing}_f(t) := [-\hat{x}_1(t-1), -\hat{x}_1(t-2), \dots, -\hat{x}_1(t-n)]^T \in \mathbb{R}^n \tag{64}$$

$$\varnothing_b(t) := [u(t-1), u(t-2), \dots, u(t-n)]^T \in \mathbb{R}^n \tag{65}$$

$$\hat{\varnothing}_v(t) := \left[\hat{v}(t-1), \hat{v}(t-2), \dots, \hat{v}(t-n_J)\right]^T \in \mathbb{R}^{n_J} \tag{66}$$

$$\hat{\gamma}(t) = \hat{\bar{w}}(t-1) + \hat{\bar{w}}(t-2) + \cdots + \hat{\bar{w}}(t-n), \tag{67}$$

$$\beta(t) := T_1 y(t-1) + T_2 y(t-2) + \cdots + T_n y(t-n) \tag{68}$$

$$\hat{\bar{F}} := \left[\hat{\bar{f}}_1, \hat{\bar{f}}_2, \dots, \hat{\bar{f}}_{\mathrm{n}}\right]^T \in \mathbb{R}^n \tag{69}$$

$$\hat{\theta} := \left[\hat{\bar{F}}^T, \hat{\bar{G}}^T, \hat{d}, \hat{J}^T\right]^T \in \mathbb{R}^{n_s}, n_s = 2n + 1 + n_J \tag{70}$$

$$xp(t+1) = \hat{\bar{F}} xp(t) + \hat{\bar{G}} u(t) + Ty(t) \tag{71}$$

$$P = \hat{\bar{F}} * P * \hat{\bar{F}}^T + Q - \hat{T} * \hat{S}^T \tag{72}$$

$$K = (P * c)/c * P * c^T + \hat{R} \tag{73}$$

$$x(t+1) = xp(t+1) + +K\widehat{v^*}(t) \tag{74}$$

$$P = (eye(n) - Kc) * P \tag{75}$$

$$\widehat{v^*}(t) = y(t) - H\hat{x}(t) - \hat{d}(t)u(t), \tag{76}$$

$$\hat{v}(t) = \widehat{v^*}(t) - \hat{J}_1\hat{v}(t-1) - \hat{J}_2\hat{v}(t-2) \tag{77}$$

$$\hat{w}(t) = \hat{x}(t+1) - \hat{\bar{F}}(t)\hat{x}(t) - \hat{\bar{G}}(t)u(t) - T\left(cx(t) + \hat{d}\,u(t)\right) \tag{78}$$

$$\hat{\bar{w}}(t) = \hat{w}(t) - T\widehat{v^*}(t) \tag{79}$$

$$\hat{x}_1(t) = x(1, t). \tag{80}$$

Now, our target system is the system described by Eq. (1)–(2)

$$x(t+1) = Fx(t) + Gu(t) + w(t)$$
$$y(t) = Hx(t) + du(t) + v^*(t)$$
$$v^*(t) := \left(1 + J_1 q^{-1} + J_2 q^{-2} + \cdots + J_{n_J} q^{-n_J}\right) v(t)$$

We substitute

$$\hat{F} = \hat{\bar{F}} + \hat{T}H$$
$$\hat{G} = \left(\hat{\bar{G}} + \hat{T}d\right)$$
$$S = \rho_{w,v}\sqrt{R}\sqrt{Q},$$
$$\overline{Q} = Q - TS^T$$

$$\hat{F} := \begin{bmatrix} -\hat{f}_1 & 1 & 0 & \cdots & 0 \\ -\hat{f}_2 & 0 & 1 & \cdots & 0 \\ \vdots & \vdots & \vdots & \ddots & \vdots \\ -\hat{f}_{n-1} & 0 & 0 & \cdots & 1 \\ -\hat{f}_n & 0 & 0 & \cdots & 0 \end{bmatrix}, \hat{G} := \begin{bmatrix} \hat{g}_1 \\ \hat{g}_2 \\ \vdots \\ \hat{g}_{n-1} \\ \hat{g}_n \end{bmatrix}.$$

Finally, we extract the system parameter estimates as:

$$\theta_{estimated} := \left[\hat{F}^T, \hat{G}^T, \hat{d}, \hat{J}^T\right]^T \in \mathbb{R}^{n_s}, n_s = 2n + 1 + n_J,$$

Remark 4: The KF-CN-RGELS algorithm demonstrates the utility of using correlation coefficients to obtain state estimates and improve the accuracy of parameter estimation. This innovative approach is pivotal in achieving accurate system estimation in cases where parameters and states are not entirely known.

7. The impact of negative correlation coefficients ρ_{wv} on state estimation accuracy

In this section, we address the impact of negative correlation coefficients on estimation accuracy. We have studied this effect by examining its influence on two fundamental factors: observation accuracy and process model accuracy.

7.1 Observation accuracy

To commence our exploration of observation accuracy, let us examine the model and measurement equations under the assumptions $F = I, H = 1$, and $u(t) = 0$ in Eqs. (31) and (32). This examination leads us to the following relationship:

$$y(t) = y(t-1) + v^*(t) + \overline{w}(t-1) + v^*(t-1) \tag{81}$$

Considering the variance of $y(t)$, we can express it as:

$$\begin{aligned} \text{var}[y(t)] &= \text{var}[y(t-1)] + \text{var}[v^*(t)] \\ &+ 2\text{cov}[y(t-1), v^*(t)] + \text{var}[\overline{w}(t-1)] \\ &+ \text{var}[v^*(t-1)] \\ &- 2\text{cov}[\overline{w}(t-1), v^*(t-1)], \end{aligned} \tag{82}$$

From (82), The measurement covariance $P(y(t))$ in term of the noise covariances $\overline{Q}(t)$, $R(t)$, and $S(t)$ can be equivalently written as

$$P(y(t)) = P(y(t-1)) + \overline{Q}(t-1) + R(t-1) - 2S(t-1) \tag{83}$$

For uncorrelated noises, i.e., $S(t) = 0$, the equation becomes:

$$P_0(y(t)) = P(y(t-1)) + \overline{Q}(t-1) + R(t-1) \tag{84}$$

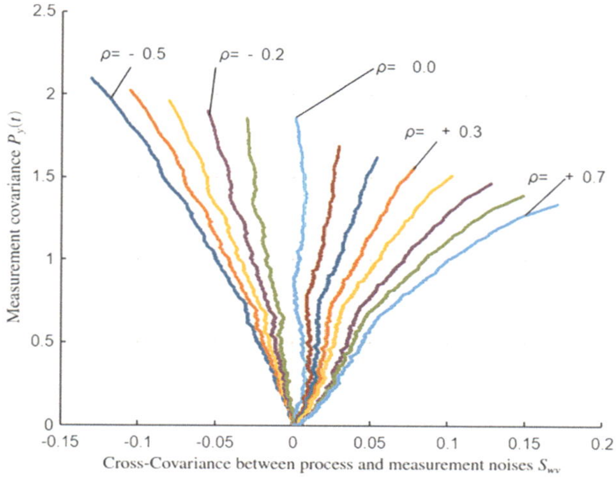

Figure 1.
Noise cross-covariance, S versus measurement covariance $P(y(t))$ for different values of correlation coefficient $\rho_{w,v}$.

According to (83) and (84), if $S(t) > 0$, then $P(y(t)) < P_0(y(t))$ indicates that the observation at time t becomes more accurate when a positive correlation coefficient exists at time $t-1$. Conversely, if $S(t) < 0$ the observation at time t becomes less accurate. It is worth noting that $\rho_{w,v} = S/\sqrt{Q}\sqrt{R}$. **Figure 1** below shows the measurement covariance $P(y(t))$ versus positive and negative values of the cross-covariance between process and measurement noises, $S(t)$. The data obtained correspond to illustrative example 1.

7.2 Process model accuracy

Turning to process model accuracy, we aim to derive the cross-covariance between the state and the process noise, defined as:

$$P(\hat{w}(k|k), \hat{x}(k|k)) = E\left[\left(w(k) - \hat{w}(k|K)\right)(x(k) - \hat{x}(k|k))^T\right] \tag{85}$$

Before we delve into the derivation, we rely on a lemma known as the Conditional Gaussian Distribution Lemma presented in Ref. [36], which proves to be invaluable.

Lemma 1: Suppose a pair of vectors Y and X are jointly Gaussian with a mean vector m_Γ and a covariance matrix $P_{\Gamma\Gamma}$, then X is conditionally Gaussian on Y with a conditional mean vector $m_{X|Y}$ and a conditional covariance matrix $P_{X|Y}$.

With this lemma, we can set:

$\Gamma = \begin{bmatrix} Y \\ X \end{bmatrix} = \begin{bmatrix} y(k) \\ w(k) \end{bmatrix} = \begin{bmatrix} Hx(k) + v(k) \\ w(k) \end{bmatrix}$, these yields

$m_\Gamma = \begin{bmatrix} \hat{y}(k|k-1) \\ \hat{w}(k|k-1) \end{bmatrix} = \begin{bmatrix} H\hat{x}(k|k-1) \\ 0 \end{bmatrix}$ And $P_{\Gamma\Gamma} = \begin{bmatrix} P_{y(k)y(k)} & P_{y(k)w(k)} \\ P_{w(k)y(k)} & P_{w(k)w(k)} \end{bmatrix}$

Where $P_{y(k)y(k)}$ is the measurement prediction covariance, $HP_{\hat{x}_{k|k-1}\hat{x}_{k|k-1}}H^T$, and $P_{w(k)y(k)}$ is the cross-covariance between process and measurement noise, S.

Hence, we can express:

$$P_{\Gamma\Gamma} = \begin{bmatrix} HP_{\hat{x}_{k|k-1}\hat{x}_{k|k-1}}H^T & S^T \\ S & Q \end{bmatrix}$$

Leveraging Lemma 1, we derive the conditional mean and covariance of the process noise.

$$\hat{w}(k|k) = \hat{w}(k|k-1) + S\left(P_{\hat{x}_{k|k-1}\hat{x}_{k|k-1}}H^T\right)^{-1}(y(k) - \hat{y}(k|k-1)) \tag{86}$$

And

$$P\left(\hat{W}(k|k) = Q - S\left(P_{\hat{x}_{k|k-1}\hat{x}_{k|k-1}}H^T\right)^{-1}S^T. \tag{87}$$

By using Eqs. (5) and (86) with (85), we can prove:

$$P(\hat{w}(k|k), \hat{x}(k|k)) = -S\left(H\,P_{\hat{x}_{k|k-1}\hat{x}_{k|k-1}}H^T\right)^{-1}HP_{\hat{x}_{k|k-1}\hat{x}_{k|k-1}} \tag{88}$$

The results presented in Eq. (88) highlight the role of the cross-covariance between the state and process noise in the context of the correlation coefficient. The correlation coefficient can be both positive or negative values and significantly impacts this cross-covariance. When the correlation coefficient is positive, increasing its value reduces the cross-covariance between the state and process noise. Conversely, when the correlation coefficient is negative, increasing its value elevates the cross-covariance between the state and process noise. **Table 1** and **Figure 2** further illustrate this outcome. The data obtained corresponds to illustrative example 1.

$\rho_{w,v}$	−0.7	−0.5	−0.3	0	0.3	0.5	0.7
Tr $[P(\hat{w}(k)|k), \hat{x}(k|k)]$	0.0166	0.0097	0.0052	−0.0001	−0.0038	−0.0059	−0.0076

Table 1.
Correlation coefficient $\rho_{w,v}$ versus trace of cross-covariance $P(\hat{w}(k|k), \hat{x}(k|k))$ between state and process noise.

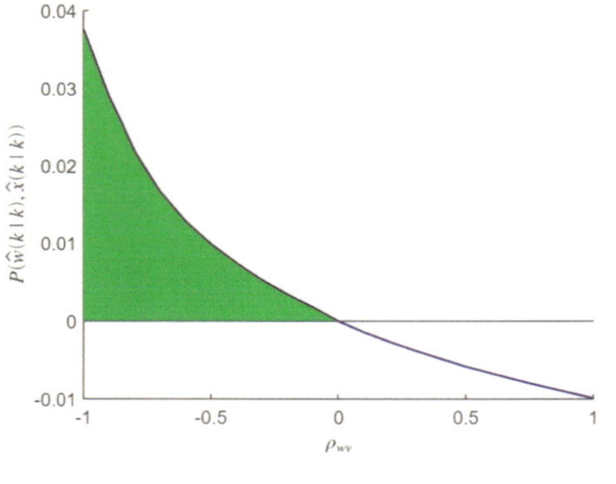

Figure 2.
Correlation coefficient $\rho_{w,v}$ versus trace of cross-covariance $P(\hat{w}(k|k), \hat{x}(k|k))$ between state and process noise.

8. Illustrative examples

Example 1. Consider the observer canonical state-space description.

$$x(t+1) = Fx(t) + Gu(t) + w(t),$$
$$y(t) = Hx(t) + du(t) + J(q)v(t).$$

$$F = \begin{bmatrix} -f_1 & 1 \\ -f_2 & 0 \end{bmatrix} = \begin{bmatrix} 0.05 & 1 \\ 0.35 & 0 \end{bmatrix}, G = \begin{bmatrix} g_1 \\ g_2 \end{bmatrix} = \begin{bmatrix} 2.0 \\ 3.0 \end{bmatrix}, H = [1, \quad 0], d = 1.30, w(t) = \begin{bmatrix} w_1(t) \\ w_2(t) \end{bmatrix}$$

$$J(q) = 1 + J_1 q^{-1} + J_2 q^{-2} = 1 + 0.0505 q^{-1} + 0.0139 q^{-2}.$$

The parameter vector to be identified is given by:

$$\begin{aligned} \theta &= [f_1, f_2, g_1, g_2, d, J_1, J_2]^T, \\ &= [-0.05, -0.35, 2.0, 3.0, 1.30, 0.0505, 0.0139]^T. \end{aligned}$$

- For reliable models, choosing system parameters must guarantee stability (avoiding unbounded oscillations), controllability (being able to reach any desired state), and observability (knowing the internal state from measurable outputs).

- In the simulation, the input $\{u(t)\}$ is a pseudo-random binary sequence generated by the MATLAB function $u =$ idinput ([65535, 1, 1], prbs' ', [0, 1], [−0.8,1]), $w_1(t)$ and $w_2(t)$ are random noise sequences with zero mean and variance $\sigma^2_{w_1} = 0.07^2$, and $\sigma^2_{w_2} = 0.01^2$ respectively. $v(t)$ is a random noise sequence with zero mean and variance $\sigma^2_v = 0.8^2$. Set the data length $L = 5000$ and choose different values of the correlation coefficient $\rho_{w,v}$ in the range [0–1]. Generate system parameter and state estimates by applying the KF-CNRGELS algorithm with different values of correlation coefficient, $\rho_{w,v}$ and examine the correlation coefficient effect.

- To facilitate simulation, the generation of correlated process and measurement noises involves creating a correlation matrix. This matrix is then applied to eigen decomposition, giving rise to a correlating filter that captures the interdependencies within the noise components.

  The parameters estimates and errors $\delta_\theta = \|\hat{\theta} - \theta\| / \|\theta\|$ at $\rho_{w,v} = 0.5, 0.6$, and 0.9 are summarized in **Table 2**.

- In **Figure 3**, we vary the correlation coefficient $\rho_{w,v}$ and plot the resulting parameter estimation errors. The plot shows three curves corresponding to correlation coefficients of 0, 0.5, and 0.9.

- The root-mean-squared error of the state estimate x_1 for $\rho_{w,v}$ in the range − 1.0 to +1.0 is shown in **Figure 4**. **Figure 5** shows the relationship between the parameter estimation error and various values of the correlation coefficient $\rho_{w,v}$ positive and negative values. **Figure 6** shows the parameter estimation error of the proposed algorithm compared to the standard Kalman filter SKF and Augmented-State Kalman filter algorithms for $\rho_{w,v} = 0.6$. **Tables 3** and **4** illustrate the results of the comparison between the KF-CN-GELS and AUG-KF for $\rho_{w,v} = 0.0$, the KF-CN-GELS and AUG-KF for $\rho_{w,v} = 0.4$ and the

KF-CN-GELS and AUG-KF for $\rho_{w,v} = 0.8$. The noise estimates $\hat{v}(t), \hat{w_1}(t)$ and $\hat{w_2}(t)$ for KF-CN-GELS with $\rho_{w,v} = 0.7$ is shown in **Figure 7**. The state estimates of $x_1(t)$and $x_2(t)$ and errors for $\rho_{w,v} = 0.7$ are depicted in **Figure 8**. The collected input and output data are shown in **Figure 9**.

$\rho_{w,v}$	t	f_1	f_2	g_1	g_2	d	J_1	J_2	δ_θ%
0.5	5000	−0.0402	−0.3557	2.0117	3.0165	1.3191	0.0580	0.0398	1.0491
0.6	5000	−0.0396	−0.3557	2.0110	3.0173	1.3185	0.0587	0.0345	0.9696
0.9	5000	0.0416	−0.3545	2.0098	3.0148	1.3120	0.0470	0.0201	0.6354
True values		0.0500	−0.3500	2.0000	3.0000	1.3000	0.0505	0.0139	

Table 2.
The KF-CN-RGELS estimates and errors $(\rho_{w,v} = 0.5, 0.6,$ *and* 0.9 *for* $(R_v = 0.8, Q_w = [0.07, 0.01]I_2)$.

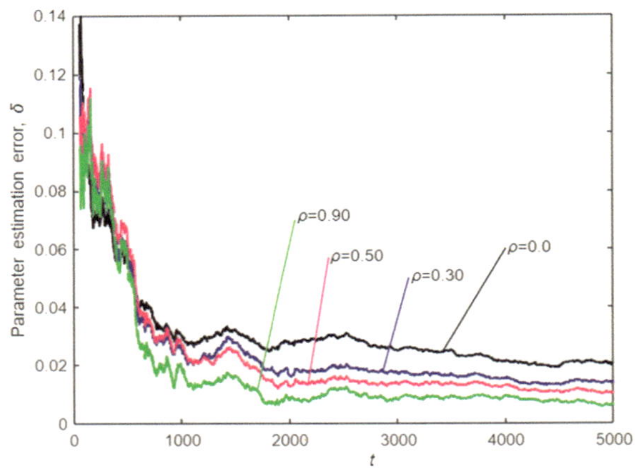

Figure 3.
The KF-CN-RGELS parameter estimation error δ_θ*. Against* $t(\rho_{w,v} = 0, 0.3, 0.5,$ *and* 0.9*, and* $R_v = 0.8, Q_w = [0.07, 0.01]I_2.)$.

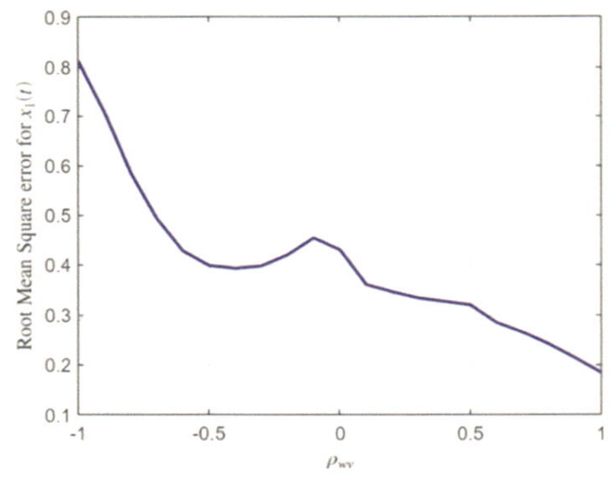

Figure 4.
Root mean square error (RMSE) for $x_1(t)$ *versus different values of correlation coefficient* $\rho_{w,v}$.

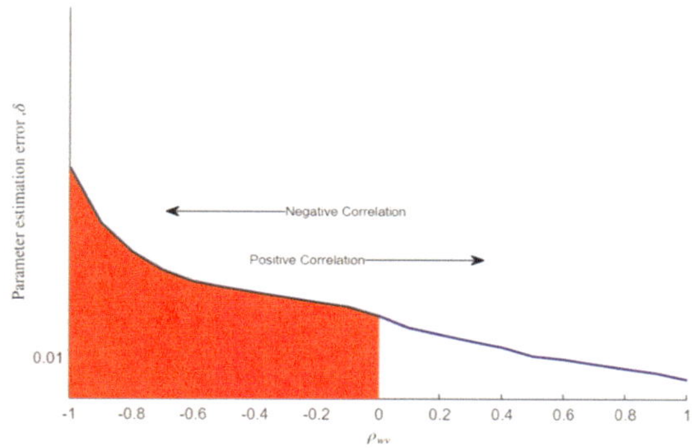

Figure 5.
The parameter estimation error δ_θ against $\rho_{w,v} = -1$ to $+1$, $(R_v = 0.15, Q_w = [0.07, 0.01]I_2)$.

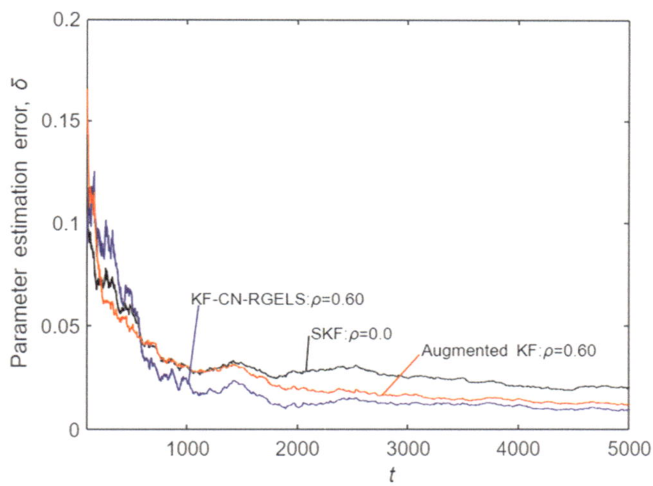

Figure 6.
The KF-CN-RGELS estimation error δ_θ against t compared to SKF and augmented KF algorithms $(\rho_{w,v} = 0.60$ for $R_v = 0.8, Q_w = [0.07, 0.01]I_2)$.

$\rho_{w,v} = 0$	t	f_1	f_2	g_1	g_2	d	J_1	J_2	$\delta_\theta\%$
SKF	5000	−0.0521	−0.3461	2.0186	3.0047	1.3108	0.0967	0.0732	2.0386
AUG-KF	5000	−0.0522	−0.3461	2.0186	3.0047	1.3108	0.0967	0.0732	2.0389
KF-CN-RGELS	5000	−0.0524	−0.3464	2.0190	3.0052	1.3108	0.0958	0.0727	2.0177
True values		0.0500	−0.3500	2.0000	3.0000	1.3000	0.0505	0.0139	

Table 3.
The parameter estimates and errors (KF-CN-RGELS, SKF, and AUG-KF) for $\rho_{wv} = 0.0$ $(R_v = 0.8, Q_w = [0.07, 0.01]I_2)$.

Algorithms	KF-CN-RGELS ($\rho_{wv} = 0.4$)	AUG-KF ($\rho_{wv} = 0.4$)	KF-CN-RGELS ($\rho_{wv} = 0.6$)	AUG-KF ($\rho_{wv} = 0.6$)	KF-CN-RGELS ($\rho_{wv} = 0.8$)	AUG-KF ($\rho_{wv} = 0.8$)
$f_1 = -0.0500$	−0.0388	−0.0413	−0.0396	−0.0446	−0.0408	−0.0515
$f_2 = -0.3500$	−0.3553	−0.3537	−0.3557	−0.3511	−0.3549	−0.3456
$g_1 = 2.0000$	2.0098	2.0103	2.0110	2.0091	2.0101	2.0090
$g_2 = 3.0000$	3.0164	3.0132	3.0173	3.0080	3.0158	2.9971
$d = 1.3000$	1.3221	1.3218	1.3185	1.3193	1.3146	1.3170
$J_1 = 0.0505$	0.0698	0.0740	0.0587	0.0570	0.0500	0.2615
$J_2 = 0.0139$	0.0452	0.0580	0.0345	0.0537	0.0256	0.1422
$\delta_\theta(\%)$	1.2622	1.5008	0.9696	1.2109	0.7413	6.4351

Table 4.
The parameter estimates and errors (KF-CN-RGELS and AUG-KF) for $(R_v = 0.8, Q_w = [0.07, 0.01]I_2)$.

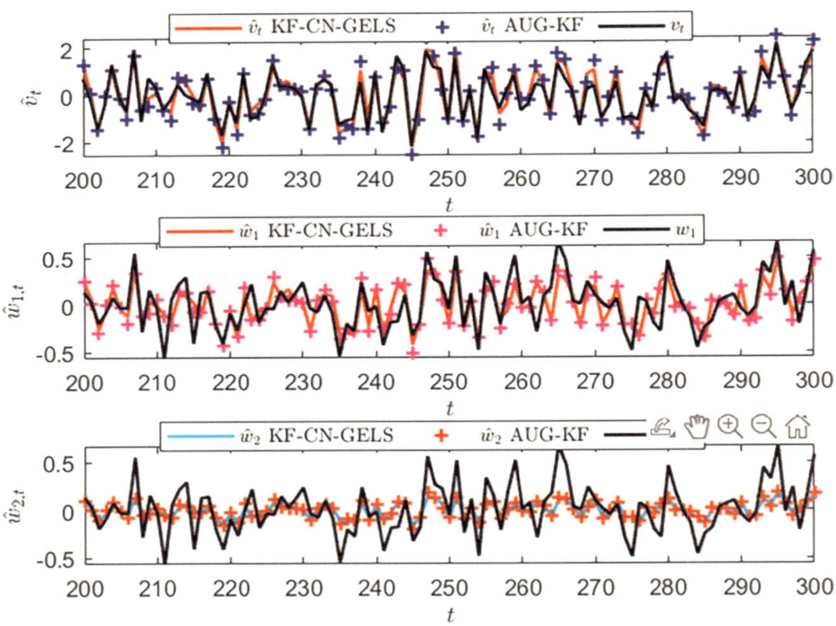

Figure 7.
The estimates of $w_2(t)$, $w_2(t)$, and $v(t)$ for KF-CN-GELS and AUG-KF $(\rho_{w,v} = 0.7$ for $R_v = 0.8, Q_w = [0.07, 0.01]I_2)$.

Looking at **Tables 2–4** and **Figures 3–9**, we can draw some conclusions from these tables and figures.

- At $\rho_{w,v} = 0$, there is no correlation between process noise and measurement noise. The filtering result of the standard Kalman filter is the same as the filtering result of the KF-CN-RGELS and the AUG-KF algorithms, as illustrated in **Table 3**.

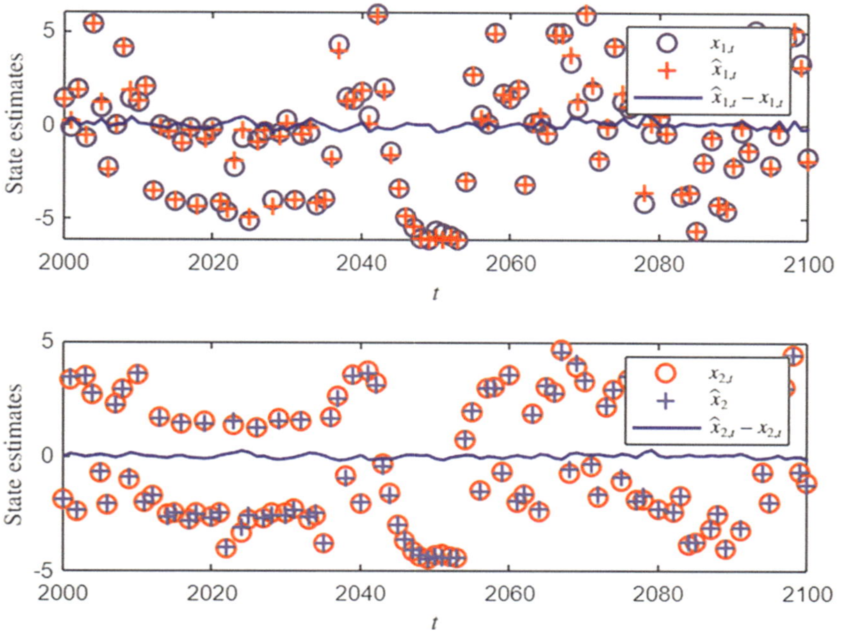

Figure 8.
The true state, estimated state, and state estimation error used in ***example 1***, $\rho_{w,v} = 0.70$.

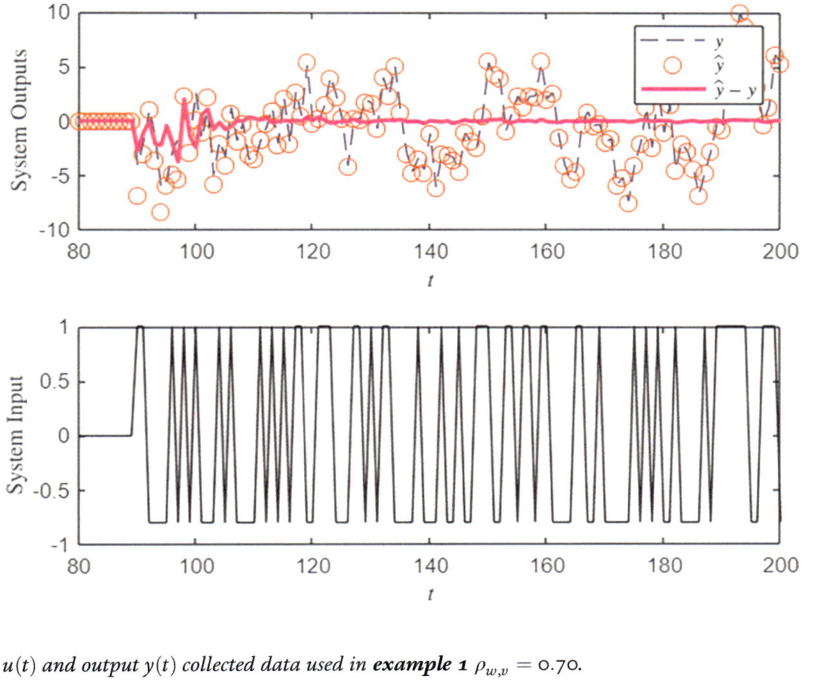

Figure 9.
The input $u(t)$ and output $y(t)$ collected data used in ***example 1*** $\rho_{w,v} = 0.70$.

- The best estimate of the state is obtained when the correlation coefficient $\rho_{w,v}$ between the process noise and the measurement noise increases. This is reflected in the improved accuracy of the parameter estimates presented by the KF-CN-RGELS algorithm. See **Figure 4** and **Table 2**.

- Clearly, the KF-CN-RGELS algorithm yields more accurate parameter estimates, since the correlation coefficient $\rho_{w,v}$ increases in the positive direction, and the mean-square error of the estimated states decreases. See **Figures 4** and **5** and **Table 2**.

- We find that the KF-CN-RGELS algorithm provides better parameter estimation accuracy than the SKF and AUG-KF algorithms under the same conditions, which makes this algorithm efficient and robust. See **Figure 6** and **Table 4**.

- The KF-CN-GELS algorithm provides better estimates of the process and measurement noises. $w_1(t)$, $w_2(t)$, and $v(t)$ than the AUG-KF algorithms under the same conditions. **Figure 7** shows these results.

Example 2. Consider the state-space model for a simple two-tank system, where $u(t)$ is the inlet water flow, $x_1(t)$ and $x_2(t)$ are the water levels of two tank, $y(t)$ is measurement of water level of tank 1 as shown in **Figure 10**. The state-space model of this system includes process and measurement noises:

$$\begin{bmatrix} x_1(t+1) \\ x_2(t+1) \end{bmatrix} = \begin{bmatrix} 0.11 & 1 \\ 0.15 & 0 \end{bmatrix} \begin{bmatrix} x_1(t) \\ x_2(t) \end{bmatrix} + \begin{bmatrix} 1.9 \\ 1.6 \end{bmatrix} u(t) + \begin{bmatrix} \omega_1(t) \\ \omega_2(t) \end{bmatrix}$$

$$y(t) = \begin{bmatrix} 1 & 0 \end{bmatrix} \begin{bmatrix} x_1(t) \\ x_2(t) \end{bmatrix} + 1.9u(t) + \left[1 + 0.1069q^{-1} - 0.0143q^{-2}\right] v(t)$$

The parameter vector to be identified is given by:

$$\begin{aligned} \theta &= \left[f_1, f_2, g_1, g_2, d, J_1, J_2\right]^T \\ &= \left[-0.11, -0.15, 1.90, 1.60, 1.90, 0.1069, -0.0143\right]^T \end{aligned}$$

The Q and R matrices are assumed to be.

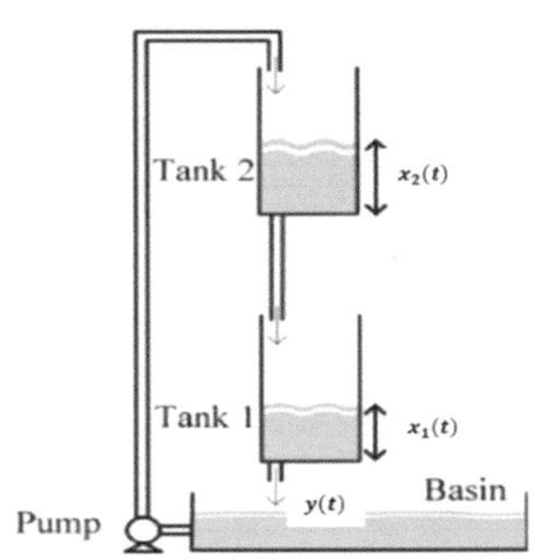

Figure 10.
The two-tank schematic diagram used in example 2.

$$Q = \begin{bmatrix} 0.60 & 0.00 \\ 0.00 & 0.40 \end{bmatrix}, R = [1.60], \text{respectively.}$$

In the case of correlated noises, we define the system noise matrix as

$$U = \begin{bmatrix} Q & S \\ S^T & R \end{bmatrix},$$

From the code for $\boldsymbol{\rho_{wv}} = \mathbf{0.95}, U = \begin{bmatrix} 0.5874 & 0.4813 & \mathbf{0.9798} \\ 0.4813 & 0.3947 & \mathbf{0.7999} \\ 0.9798 & 0.7999 & 1.7013 \end{bmatrix}$

The theoretical value of S is:

$S = \rho_{wv}\sqrt{Q}\sqrt{R} = 0.95 \times \sqrt{[0.6; 0.4]} \times \sqrt{1.6} = \begin{bmatrix} \mathbf{0.9308} \\ \mathbf{0.7600} \end{bmatrix}$, which is approximately $\begin{bmatrix} \boldsymbol{U}(\mathbf{1}, \mathbf{3}) \\ \boldsymbol{U}(\mathbf{2}, \mathbf{3}) \end{bmatrix}$.

During simulation, the input and output data that are collected from a pseudo-random binary sequence denoted by $u(t)$ and the water level measurement of tank1 $y(t) = x_1(t)$ are shown in **Figure 10**, $w_1(t)$ and $w_2(t)$ are white noise sequences with zero mean and variances $\sigma_{w_1}^2 = 0.60^2$ and $\sigma_{w_2}^2 = 0.40^2$ respectively. $v(t)$ is a white noise sequence with zero mean and variance $\sigma_v^2 = 1.60^2$.Set the data length $L = 5000$. Apply the KF-CN-RGELS algorithm to identify this two-tank model. To test the performance of the algorithm, different values of correlation coefficient (positive and negative values) are used for the system, and the simulation results are displayed in **Tables 5–8** and **Figure 11**.

When choosing the correlation coefficient for the simulation, it is important to make sure that the noise matrix U above is positive semidefinite, which means that all its eigenvalues are positive and the eigenvectors are orthogonal.

$\rho_{w,v}$	t	f_1	f_2	g_1	g_2	d	J_1	J_2	$\delta_\theta\%$
0.53	5000	−0.1293	−0.1322	1.8892	1.5373	1.9070	0.1536	−0.0914	3.6241
0.75	5000	−0.0960	−0.1591	1.8915	1.6051	1.9030	0.1263	−0.0990	2.8425
True values		−0.1100	−0.1500	1.9000	1.6000	1.9000	0.1069	−0.0143	

Table 5.
The KF-CN-RGELS estimates and errors $(\rho_{w,v} = 0.53$ *and* 0.75 *for* $(R_v = 1.6, Q_w = [0.6, 0.4]I_2)$.

$\rho_{w,v} = 0$	t	f_1	f_2	g_1	g_2	d	J_1	J_2	$\delta_\theta\%$
SKF	5000	−0.1213	−0.1418	1.8904	1.5702	1.9004	0.3480	0.0832	8.3683
AUG-KF	5000	−0.1216	−0.1418	1.8903	1.5693	1.9003	0.3360	0.0721	7.8924
KF-CN-RGELS	5000	−0.1221	−0.1413	1.8901	1.5676	1.9003	0.3301	0.0607	7.6053
True values		−0.1100	−0.1500	1.9000	1.6000	1.9000	0.1069	−0.0143	

Table 6.
The parameter estimates and errors (KF-CN-RGELS, SKF, and AUG-KF) for $\rho_{wv} = 0.0$ $(R_v = 1.6, Q_w = [0.6, 0.4]I_2)$.

Algorithms	KF-CN-RGELS ($\rho_{wv}=0.3$)	AUG-KF ($\rho_{wv}=0.3$)	KF-CN-RGELS ($\rho_{wv}=0.5$)	AUG-KF ($\rho_{wv}=0.5$)	KF-CN-RGELS ($\rho_{wv}=0.7$)	AUG-KF ($\rho_{wv}=0.7$)
$f_1=-0.1100$	−0.1487	−0.1242	−0.1336	−0.0997	−0.1019	−0.0932
$f_2=-0.1500$	−0.1181	−0.1424	−0.1286	−0.1588	−0.1546	−0.1600
$g_1=1.9000$	1.8897	1.9012	1.8888	1.8969	1.8914	1.8957
$g_2=1.6000$	1.5047	1.5695	1.5286	1.6083	1.5934	1.6161
$d=1.9000$	1.9103	1.9097	1.9076	1.9062	1.9037	1.9057
$J_1=0.1069$	0.2016	0.3822	0.1586	0.2400	0.1316	0.2045
$J_2=-0.0143$	−0.0409	0.1490	−0.0871	0.0201	−0.1004	−0.0001
$\delta_\theta(\%)$	4.6767	10.2745	3.8111	4.4205	2.8971	3.2575

Table 7.
The parameter estimates and errors (KF-CN-RGELS and AUG-KF) for $(R_v = 1.6, Q_w = [0.6, 0.4]I_2)$.

Algorithms	t	f_1	f_2	g_1	g_2	d	J_1	J_2	$\delta_\theta\%$
$KF-CN-RGELS_{\rho_{wv}=-0.5}$	5000	−0.0241	−0.1249	1.9996	1.8662	1.9901	0.4607	0.3575	19.1496
$AUG-KF_{\rho_{wv}=-0.5}$	5000	−0.1245	−0.1432	1.8957	1.5831	1.8937	0.3581	0.0998	8.8369
True values		−0.1100	−0.1500	1.9000	1.6000	1.9000	0.1069	−0.0143	

Table 8.
The parameter estimates and errors (KF-CN-RGELS and AUG-KF) for negative values of correlation coefficient $\rho_{wv} = -0.5$ $(R_v = 1.6, Q_w = [0.6, 0.4]I_2)$.

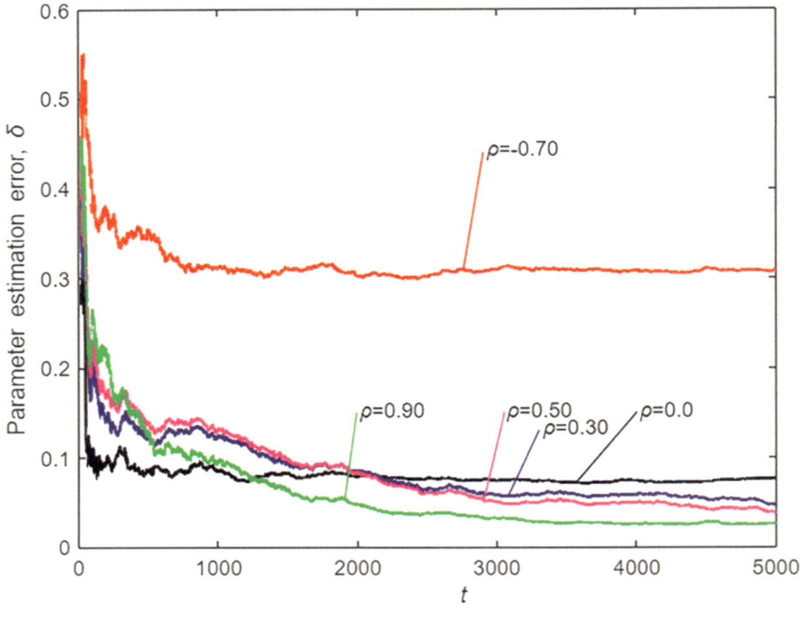

Figure 11.
The KF-CN-RGELS parameter estimation error δ_θ.

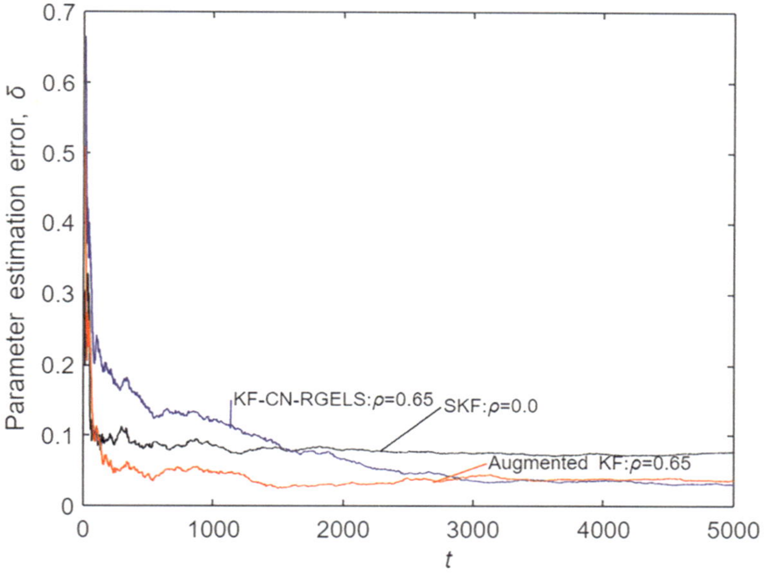

Figure 12.
The KF-CN-RGELS estimation error δ_θ against t compared to SKF.

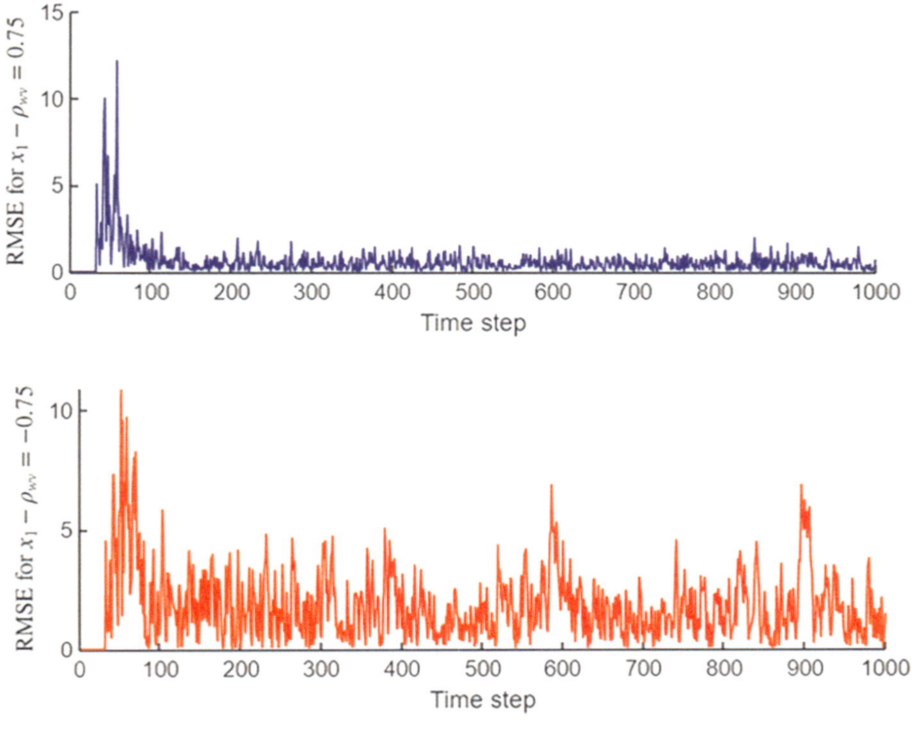

Figure 13.
The root mean square error of the state x1 versus time.

Concentrating on **Tables 5–8** and **Figures 11–14**, we can draw some conclusions from the table and figures.

- The parameter estimation errors produced by the KF-CN-RGELS algorithm decrease as the correlation coefficient increases in the positive direction. See **Figure 11** and **Table 5**.
- Under the same data length and the same correlation coefficient ρ_{wv} the KF-CN-RGELS algorithm has a faster convergence rate than the SKF and AUG-KF algorithms – see **Figure 12** and **Tables 6** and 7.
- When the correlation coefficient is negative, the model and observations are less accurate. This affects the accuracy of the state estimation, which in turn affects the accuracy of the parameter estimation – see **Table 8** and **Figure 11** refer to Section 7 for details).
- The type of correlation between the process and measurement noises (increasing positively or negatively) affects the root mean-quare error of the system state. A highly positive correlation reduces the root mean square error, whereas a highly negative correlation increases the root mean square error of system states; see **Figure 13**.

Against $t(\rho_{w,v} = 0, 0.3, 0.5, 0.9, \text{and} - 0.7, \text{for } R_v = 1.6, Q_w = [0.6{,}0.4]I_2)$ and Augmented KF algorithms $(\rho_{w,v} = 0.65 \text{ for } R_v = 1.6, Q_w = [0.6{,}0.4]I_2)$

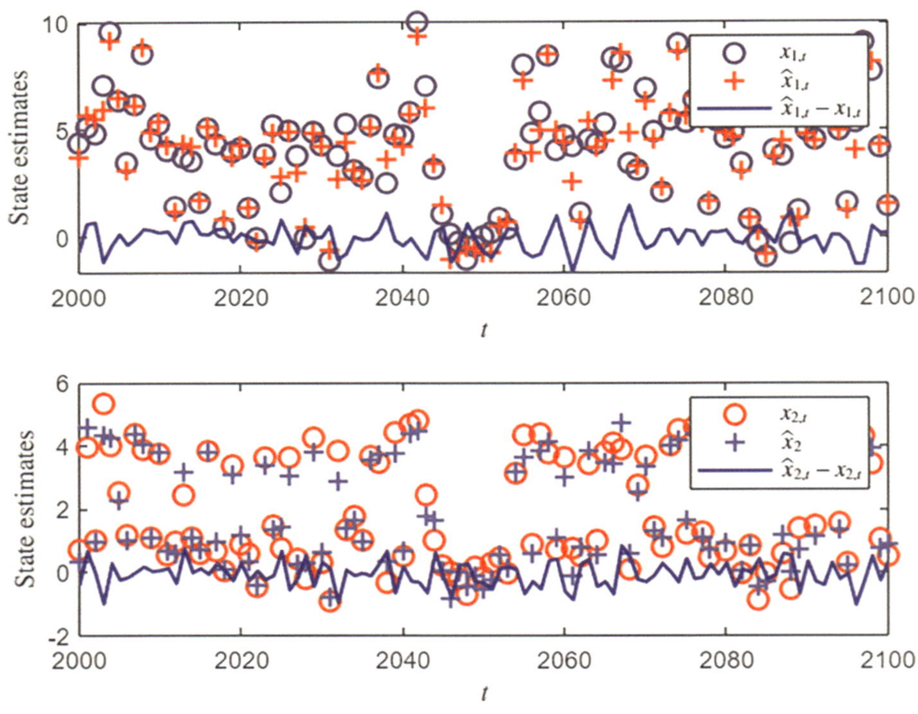

Figure 14.
The true state, estimated state and state estimation error used in example 1, for KF – CN – RGELS $(\rho_{w,v} = (0.75, -0.75)$ *for* $R_v = 1.6, Q_w = [0.6{,}0.4]I_2)$ $\rho_{w,v} = 0.75$.

9. Conclusion

The papers in this issue propose the KF-CN-RGELS algorithm to solve the combined state and parameter estimation problem in stochastic linear state-space systems perturbed by correlated processes and measurement noise. In this study, a known degree of correlation between measurement noise and process noise was theoretically predefined. To estimate the state of the system, the values of the correlation coefficients were chosen based on the considered real-time application situation. We developed a variant of the Kalman filter that considers noise correlations by adjusting the process noise matrix. Simulations show this new approach is successful, improving parameter estimation accuracy as noise correlation becomes more positive. The effectiveness of the proposed algorithm was tested theoretically by selecting several values for the positive and negative correlation coefficient. A negative correlation coefficient can have negative effects on the accuracy of both observations and models. It causes the measurements to be less reliable and introduces more uncertainty in determining the relationship between the observed data and the underlying process. This uncertainty, in turn, affects the filtering process that is used to estimate the true state and parameters from the observed data. As a result, the state and parameter estimates become less accurate. Many researchers in literature provide efficient ways to estimate the covariance matrices Q, R, and S that will be used in future work to guarantee best practice results. By estimating these matrices, the optimal correlation coefficient can be practically chosen, and according to Eq. (42), the algorithm gives satisfactory results from a practical point of view. Beyond its application to linear stochastic systems with KF-CN-RGELS, this approach opens doors to developing novel algorithms and tackling the identification of more complex systems like bilinear ones. Furthermore, the proposed KFCN-RGELS can be used to study real-time applications such as aircraft radar guidance systems, which is the case for cross-correlated noise systems.

Conflict of interest

The authors declare no conflict of interest.

Author details

Abd El Mageed Hag Elamin Khalid
College of Engineering, Al Neelain University, Khartoum, Sudan

*Address all correspondence to: elaminkhalid32@gmail.com

References

[1] Tewari A. Modern Control Design with MATLAB and SIMULINK. Vol. 1. Chichester: Wiley; 2002

[2] Mu B, Bai E-W, Zheng WX, Zhu Q. A globally consistent nonlinear least squares estimator for identification of nonlinear rational systems. Automatica. 2017;**77**:322-335

[3] Ding F, Xu L, Meng D, Jin XB, Alsaedi A, Hayat T. Gradient estimation algorithms for the parameter identification of bilinear systems using the auxiliary model. Journal of Computational and Applied Mathematics. 2020;**369**:112575

[4] Krishnanathan K, Anderson SR, Billings SA, Kadirkamanathan V. Computational system identification of continuous-time nonlinear systems using approximate Bayesian computation. International Journal of Systems Science. 2016;**47**(15):3537-3544

[5] Pan W, Yuan Y, Gonsalves J, Stan G-B. A sparse Bayesian approach to the identification of nonlinear state-space systems. IEEE Transactions on Automatic Control. 2015;**61**(1):182-187

[6] Gan M, Chen CLP, Chen G-Y, Chen L. On some separated algorithms for separable nonlinear least squares problems. IEEE Transactions on Cybernetics. 2017;**48**(10):2866-2874

[7] Gan M, Li H-X. An efficient variable projection formulation for separable nonlinear least squares problems. IEEE Transactions on Cybernetics. 2013; **44**(5):707-711

[8] Gu Y, Liu J, Li X, Chou Y, Ji Y. State space model identification of multirate processes with time-delay using the expectation maximization. Journal of the Franklin Institute. 2019;**356**(3): 1623-1639

[9] Li B. State estimation with partially observed inputs: A unified Kalman filtering approach. Automatica. 2013; **49**(3):816-820

[10] Gil P, Henriques J, Cardoso A, Dourado A. On affine state-space neural networks for system identification: Global stability conditions and complexity management. Control Engineering Practice. 2013;**21**(4):518-529

[11] Zhang X et al. Combined state and parameter estimation for a bilinear state space system with moving average noise. Journal of the Franklin Institute. 2018; **355**(6):3079-3103

[12] Safarinejadian B, Asad M, Sadeghi MS. Simultaneous state estimation and parameter identification in linear fractional order systems using colored measurement noise. International Journal of Control. 2016; **89**(11):2277-2296

[13] Yu C, Ljung L, Wills A, Verhaegen M. Constrained subspace method for the identification of structured state-space models (COSMOS). IEEE Transactions on Automatic Control. 2019;**65**(10): 4201-4214

[14] Yu C, Ljung L, Verhaegen M. Identification of structured state-space models. Automatica. 2018; **90**:54-61

[15] Razmjooei H, Safarinejadian B. A novel algorithm for hierarchical state and parameter estimation in slowly time varying systems. Journal of Advanced and Applied Sciences (JAAS). 2015;**3**(5): 189-200

[16] Li M, Liu X, Ding F. The filtering-based maximum likelihood iterative estimation algorithms for a special class of nonlinear systems with autoregressive moving average noise using the hierarchical identification principle. International Journal of Adaptive Control and Signal Processing. 2019; **33**(7):1189-1211

[17] Cui T, Ding F, Alsaedi A, Hayat T. Data filtering-based parameter and state estimation algorithms for state-space systems disturbed by colored noises. International Journal of Systems Science. 2020;**51**(9):1669-1684

[18] Wang X, Ding F, Alsaedi A, Hayat T. Filtering based parameter estimation for observer canonical state space systems with colored noise. Journal of the Franklin Institute. 2017;**354**(1):593-609

[19] Wang Y, Ding F. Filtering-based iterative identification for multivariable systems. IET Control Theory & Applications. 2016;**10**(8):894-902

[20] Ma X, Ding F. Gradient-based parameter identification algorithms for observer canonical state space systems using state estimates. Circuits, Systems, and Signal Processing. 2015;**34**(5): 1697-1709

[21] Cui T, Ding F, Li X, Hayat T. Kalman filtering based gradient estimation algorithms for observer canonical state-space systems with moving average noises. Journal of the Franklin Institute. 2019;**356**(10):5485-5502

[22] Cui T, Ding F, Jin X-B, Alsaedi A, Hayat T. Joint multi-innovation recursive extended least squares parameter and state estimation for a class of state-space systems. International Journal of Control, Automation and Systems. 2020;**18**(6): 14121424

[23] Xu L, Ding F. Parameter estimation algorithms for dynamical response signals based on the multi-innovation theory and the hierarchical principle. IET Signal Processing. 2017;**11**(2): 228-237

[24] Baker RC, Charlie B. Nonlinear unstable systems. International Journal of Control. 1989;**23**(4):123-145

[25] Xu R, Ding F. Parameter estimation for control systems based on impulse responses. International Journal of Control, Automation and Systems. 2017; **15**(6):2471-2479

[26] Zhang X, Ding F. Adaptive parameter estimation for a general dynamical system with unknown states. International Journal of Robust and Nonlinear Control. 2020;**30**(4):1351-1372

[27] Zhang X, Ding F. Hierarchical parameter and state estimation for bilinear systems. International Journal of Systems Science. 2020;**51**(2):275-290

[28] Zhang X, Ding F. Recursive parameter estimation and its convergence for bilinear systems. IET Control Theory & Applications. 2020; **14**(5):677-688

[29] ElAmin H, El Mageed KA. Clustering input signals based identification algorithms for two-input single-output models with autoregressive moving average noises. Complexity. 2020;**2020**: 1-12

[30] Elamin KAEMH. State estimation on correlated noise and unit time-delay systems. In: 2016 Conference of Basic Sciences and Engineering Studies (SGCAC). Khartoum, Sudan; 2016. pp. 94-100

[31] Elamin KAEMH, Taha MFE. On the steady-state error covariance matrix of

Kalman filtering with intermittent observations in the presence of correlated noises at the same time. In: 2013 International Conference on Computing, Electrical and Electronic Engineering (ICCEEE). Khartoum, Sudan; 2013. pp. 15-22

[32] Jiang P, Zhou J, Zhu Y. Globally optimal Kalman filtering with finite-time correlated noises. In: 49th IEEE Conference on Decision and Control (CDC). Atlanta, GA, USA; 2010. pp. 5007-5012

[33] Wang X, Liang Y, Pan Q, Yang F. A Gaussian approximation recursive filter for nonlinear systems with correlated noises. Automatica. 2012;**48**(9): 2290-2297

[34] Wang X, Liang Y, Pan Q, Wang Z. General equivalence between two kinds of noisecorrelation filters. Automatica. 2014;**50**(12):3316-3318

[35] Chang G. Alternative formulation of the Kalman filter for correlated process and observation noise. IET Science, Measurement & Technology. 2014;**8**(5): 310-318

[36] Chang G. Marginal unscented Kalman filter for cross-correlated process and observation noise at the same epoch. IET Radar, Sonar and Navigation. 2014;**8**(1):54-64

Chapter 8

Application of the Kalman Filter in Monitoring, Diagnosis, and Fault Parrying Problems for Observable Dynamical Systems

Alexander Chernodarov

Abstract

This section is devoted to approaches for the formation, estimation and application of diagnostic parameters, as well as stochastic models in monitoring problems. The procedures for increasing the reliability and depth of dynamic systems monitoring are substantiated based on the decomposition of diagnostic models, the use of combined statistical criteria and the processing of observations in "forward" and "backward" time using the extended Kalman filter. The possibility of detecting and parrying anomalous observations using robust filtering procedures is shown. Such procedures are based on the use of an influence function that establishes the level of confidence for formed observations. *U-D* modification of the Kalman filter with an influence function in the observation selection loop is presented. Here *U* is an upper triangular matrix with unit diagonal elements and *D* is a diagonal matrix. Results of the mathematical simulation are given.

Keywords: dynamic system, monitoring, diagnosis, Kalman filter, consent criteria, decisive rules

1. Introduction

At present, the problem of increasing the reliability of dynamical systems (DSs), which are part of objects of various physical natures, still remains topical. The solution to the above problem can be based on the application of analytical approaches to fault detection and isolation (FDI). In observable dynamic systems, the FDI problem is solved using external information, diagnostic models, and decision rules. When constructing diagnostic models, a mathematical description of the reference (unperturbed) DS and the actual (perturbed) DS is used. The reference vector $Y(t)$ and the actual vector $Y_a(t)$ of state parameters correspond to such DSs. The dynamics of changes in these vectors are described by the following differential equations:

$$dY(t)/dt = \dot{Y}(t) = F[Y(t)], \tag{1}$$

$$\dot{Y}_a(t) = F[Y_a(t)] + G(t)\xi(t), \tag{2}$$

IntechOpen

where $F(\dots)$ is a matrix of variable coefficients characterizing the dynamics of changes in DS parameters; $\xi(t)$ is a vector of random deviations of DS parameters from the required values; $E\left[\xi(t)\xi^T(t-\tau)\right] = Q(t)\delta(t-\tau)$ is the covariance matrix; $Q(t)$ is a matrix of weighting coefficients; $\delta(t-\tau)$ is the delta-function; $E\,[\dots]$ is the operator of mathematical expectation; $G(t)$ is a matrix of variable coefficients characterizing the dynamics of changes in random deviations.

Taking into account relations (1) and (2), the equation for DS errors will have the form

$$dx(t)/dt = \dot{x}(t) = A(t)x(t) + G(t)\xi(t), \tag{3}$$

where $x(t) = \Delta Y(t) = Y_a(t) - Y(t)$ is the vector of DS errors; $A(t) = \partial F[Y(t)]/\partial Y|_{Y(t)=Y_a(t)}$ is a matrix of variable coefficients characterizing the dynamics of changes in DS errors.

The estimates $\hat{x}(t)$ of DS errors can be obtained by processing observations $z(t)$ using the extended Kalman filter (EKF) [1]:

$$z(t) = h[Y_a(t)] - h[Y(t)]_{\text{SEI}}, \tag{4}$$

where SEI is the sensor of external information; $h[Y(t)]_{\text{SEI}}$ is the observable parameters formed by SEI and having a model $h[Y(t)]_{\text{SEI}} = h[Y(t)] + \vartheta(t)$; and $\vartheta(t)$ is a vector of random deviations of SEI parameters from the reference with covariance matrix $E\left[\vartheta(t)\vartheta^{\mathrm{T}}(t-\tau)\right] = R(t)\delta(t-\tau)$.

The mathematical model of observation (4) has the form:

$$z(t) = H(t)x(t) + \vartheta(t), \tag{5}$$

where $H(t) = \partial h[Y(t)]/\partial Y|_{Y(t)=Y_a(t)}$ is the matrix for the relation of observed parameters and the vector of DS errors.

The extension of the traditional linear Kalman filter is associated with its application in optimization problems of nonlinear dynamic systems of the form (2)–(5). The observable DS with EKF in the errors estimation loop is shown in **Figure 1**. $\hat{Z}$ is the predicted estimate of observations; ^ is the symbol for estimate; and ν is the vector of residuals. Such a vector can be used to form diagnostic parameters for monitoring observed dynamical systems with EKF in the fault detection loop.

Discrete observations $Z_i = Z(t_i)$ are related with the following DS error equation:

$$x_i = \Phi_i x_{i-1} + \Gamma_i \xi_{i-1}, \tag{6}$$

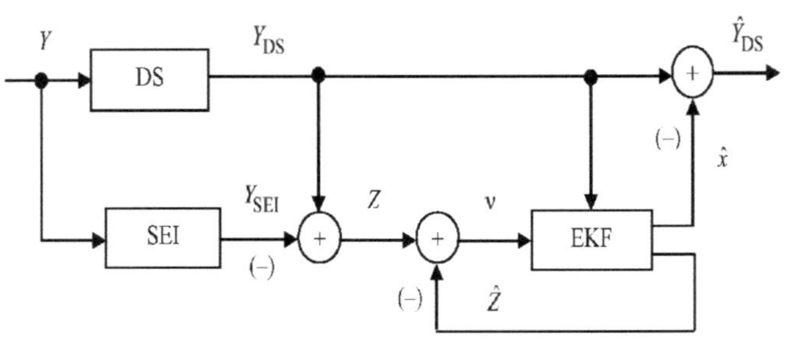

Figure 1.
An observable dynamical system with the EKF in an error estimation loop.

where $x_i = x(t_i)$; Φ_i matrix is the solution of the homogeneous part of Eq. (3).

$$d\Phi(t, t_{i-1})/dt = \dot{\Phi}_i = A(t)\Phi(t, t_{i-1}); \quad \Phi(t_{i-1}, t_{i-1}) = I;$$

I is an identity matrix; Γ_i matrix is the solution of the inhomogeneous part of Eq. (3):

$$d\Gamma(t, t_i)/dt = \dot{\Gamma}_i = A(t)\Gamma(t, t_{i-1}) + G(t); \; \Gamma(t_{i-1}, t_{i-1}) = 0.$$

The use of EKF for FDI problem can be based on the estimation of DS errors. However, in this case, it is necessary to associate each DS state with its own equation of the form (3) and its own filter. There is a need to create a bank of filters [2], which is difficult to implement in practice. Therefore, it is advisable to form FDI algorithms based on the serviceable states of the DS. Then, the diagnostic parameters will reflect the deviation of the actual state of the DS from the serviceable one.

2. Fault detection by the χ^2 Chi-Square criterion

The functioning of the EKF is based on the use of a vector of residuals (see **Figure 1**)

$$\nu_i = Z_i - \hat{Z}_i = \left[\nu_{1(i)}\nu_{2(i)} \ldots \nu_{j(i)} \ldots \nu_{l(i)}\right]^{\mathrm{T}}, \tag{7}$$

where $\hat{Z}_i = H_i\hat{x}_{i/i-1}$ and $\hat{x}_{i/i-1} = \Phi_i\hat{x}_{i-1/i-1}$ is the predicted values of estimates of the DS errors at the instant $t = t_i$ of time after Z_{i-1} observations have been processed.

The statistical properties of the residuals make it possible to form diagnostic parameters based on them. It is known [2] that the serviceable state of the DS and acceptable error values correspond to a vector of residuals that has a Gaussian distribution with zero mathematical expectation and a covariance matrix α_i, that is,

$$\nu_i \in N\left(0, \alpha_i^2\right). \tag{8}$$

Taking into account the principle of orthogonality of optimal estimates [1] $E\left[e_i\vartheta_i^{\mathrm{T}}\right] = 0$, it can be shown that

$$\alpha_i^2 = E\left[\nu_i\nu_i^T\right] = H_iP_{i/i-1}H_i^T + R_i, \tag{9}$$

where $P_{i/i-1} = E\left[e_{i/i-1}e_{i/i-1}^T\right] = \Phi_iP_{i-1/i-1}\Phi_i^T + \Gamma_iQ_{i-1}\Gamma_i^T$ is the predicted covariance matrix; $e_{i/i-1} = x_i - \hat{x}_{i/i-1}$.

In traditional EKF, all elements of the ν_i vector are processed simultaneously. Taking into account such processing, to monitor the state of the DS, it is necessary to check the l-dimensional vector of residuals for Gaussianity. In practice, the solution of this problem is a matter of some difficulty. Therefore, using the residual vector, more compact diagnostic parameters are formed. One of these parameters is formed by convolving the vector (7) and the covariance matrix (9)

$$J_i = \nu_i^T\alpha_i^{-2}\nu_i. \tag{10}$$

In the quadratic form (10), the elements of the α_i matrix are the normalizing coefficients that take into account the required statistical characteristics of the residuals. It can be shown [2] that if the ν_i vector has a Gaussian distribution, then the quadratic form (10) is distributed according to the χ^2 law with l degrees of freedom, namely,

$$J_i \in \chi^2(l,\ 2l), \tag{11}$$

that is, the l dimension of the ν_i vector is equal to the mathematical expectation of the J_i parameter and to one-half of its variance.

Relation (11) reflects the necessary condition for the absence of violations in the DS. Quadratic form (10) includes all l components of the ν_i vector, so it can be considered as a DS generalized state parameter. It is believed that if the J_i parameter is within acceptable limits, then there are no violations in the DS. The tolerances for the J_i parameter are formed taking into account the numerical characteristics of the χ^2 distribution and a given level of significance for the goodness-of-fit criterion. Such conditions can be performed taking into account the properties of the $a(l)$ quantile for the χ^2 distribution, namely,

$$P\left\{J_i > J_{T,\ a(l)}\right\}, \tag{12}$$

where $P\{\ldots\}$ is a probability measure; $J_{T,\ a(l)}$ is a tabular value of the J_i parameter for the quantile $a(l)$ with the number l of degrees of freedom; $a(l)$ is the quantile of a order, $a \in (0, 1)$.

An analysis of the tabular data shows that quantile (12) with highly accuracy reflects the three mean square deviations rule (the 3σ rule). This rule finds application in the processing of scalar parameters ξ distributed according to the Gauss law and boils down to the following. The probabilities of deviations of such parameters from the mathematical expectation are characterized by the following values [3]:

$$P\{|\xi\text{-}E[\xi]| \geq k\sigma\} = \begin{cases} 0.3173\ldots, & k = 1; \\ 0.0455\ldots, & k = 2; \\ 0.0027\ldots, & k = 3. \end{cases} \tag{13}$$

For k = 3, relation (13) reflects the 3σ rule for the Gaussian distribution law. Taking into account the probabilistic characteristics (13) and the 0.02(l) quantile of [4], it can be argued that, with a confidence probability of 0.98, the necessary condition for the J_i parameter to belong to the χ^2 distribution is associated with the fulfillment of the following inequality:

$$J_i \leq \gamma_l^2 = E[J_i] + 3\sqrt{V[J_i]} = l + 3\sqrt{2l}, \tag{14}$$

where $V[\ldots]$ is an operator for variance.

Taking into account the γ_l^2 tolerance, the monitoring of DS using the diagnostic parameter J_i for the χ^2 criterion is reduced to checking the following conditions:

$$\left.\begin{array}{l} \text{if } J_i \leq \gamma_l^2,\ \text{there are no faults in the DS;} \\ \text{if } J_i > \gamma_l^2,\ \text{there are faults in the DS.} \end{array}\right\} \tag{15}$$

Monitoring using parameter (10) allows you to determine the state of the DS as a whole without diagnosing which element of the observation vector the violation is associated with. This approach is traditional [5, 6] for DS monitoring. In practice, there is a need to diagnose DS with a depth up to the element of the observation vector.

3. Fault diagnosis by the χ^2 Chi-Square criterion

The problem of DS diagnosis with a depth of up to elements of the vector Z_{i-1} can be solved if the observation errors ϑ_i are statistically independent (uncorrelated), that is, if the matrix R_i in relation (9) is diagonal. If the observation errors are mutually correlated, their preliminary decomposition can be performed. Such a decomposition can be performed taking into account the structure of the covariance R_i. As the specified matrix is square, positive definite and symmetric, it can be represented in the following form [7]:

$$R_i = S_i D_i S_i^T, \tag{16}$$

where S_i is an upper triangular matrix with unit diagonal elements and D_i is a diagonal matrix.

Then, the vector of observations $S_i^{-1} Z_i$ will have elements with mutually uncorrelated noise. Indeed, the following model corresponds to such observations:

$$S_i^{-1} Z_i = S_i^{-1} H_i x_i + S_i^{-1} \vartheta_i \tag{17}$$

or

$$\tilde{Z}_i = \tilde{H}_i x_i + \tilde{\vartheta}_i, \tag{18}$$

where

$$\tilde{Z}_i = S_i^{-1} Z_i; \qquad \tilde{H}_i = S_i^{-1} H_i; \qquad \tilde{\vartheta}_i = S_i^{-1} \vartheta_i.$$

The covariance matrix has the form:

$$\tilde{R}_i = E\left[\tilde{\vartheta}_i \tilde{\vartheta}_k^T\right] = S_i^{-1} E\left[\vartheta_i \vartheta_k^T\right] S_i^{-T} = S_i^{-1} R_i S_i^{-T} = S_i^{-1} S_i D_i S_i^T S_i^{-T} = D_i, \tag{19}$$

where $D_i = diag\{D_{1(i)}, \ldots D_{j(i)}, \ldots D_{m(i)}\}$; $S_i^{-T} = \left(S_i^{-1}\right)^T$.

Expression (19) shows that $\tilde{Z}_i$ observations have uncorrelated errors. Next, we will assume that the matrix R_i is diagonal. Then, such observations can be processed using a sequential modification of the EKF [8].

Prediction:

$$m_0 = \hat{x}_{i/i-1} = \Phi_i \hat{x}_{i-1/i-1}; \tag{20}$$

$$M_0 = P_{i/i-1} = \Phi_i P_{i-1/i-1} \Phi_i^T + \Gamma_i Q_{i-1} \Gamma_i^T. \tag{21}$$

Updating:

$$\nu_j = z_j - H_j m_{j-1}; \tag{22}$$

$$\alpha_j^2 = H_j M_{j-1} H_j^T + R_j; \tag{23}$$

$$K_j = M_{j-1} H_j^T / \alpha_j^2; \tag{24}$$

$$M_j = (K_j H_j - I) M_{j-1} (K_j H_j - I)^T + K_j R_j K_j^T; j = \overline{1, l}; \tag{25}$$

$$\hat{x}_{i/i} = m_l; \qquad P_{i/i} = M_l, \tag{26}$$

where $\hat{x}_{i/i-1}$ and $\hat{x}_{i/i}$, respectively, are the predicted and updated estimates of the DS error vector at the ith time point, obtained after processing observations Z_{i-1} and Z_i;

$P_{i/i-1}$ and $P_{i/i}$ covariance matrices of these estimates; m_j and M_j, respectively, estimates of the DS error vector and their covariance matrix obtained after processing the jth observation; $H_i = \left[H_1^T \dots H_j^T \dots H_l^T\right]^T$ is the matrix for the relation of observed parameters with the DS error vector; H_j is the row vector for the relation of the z_j observation with the DS errors vector; R_j is the variance of the z_j observation error; Q_i is a matrix of weighting coefficients characterizing the noise intensities.

Unlike the traditional EKF, in algorithm (22)–(26), when calculating the amplification coefficient (24), it is not necessary to invert the covariance matrix (9) for the residual vector (7). Such inversion is replaced by l division operations in Eq. (24). In addition, based on the presented algorithm, it seems possible to form statistical criteria of agreement for scalar diagnostic parameters.

When processing the residual vector (7) element-by-element, algorithm (22)–(26) can be supplemented with procedures for analyzing the DS state for each of the l observations. For example, the checking of DS on the jth observation can be performed using a normalized residual $\beta_j = \nu_j / \alpha_j$. The statistical properties of such residuals make it possible to form scalar goodness-of-fit criteria. By analogy with the diagnostic parameter (10), for $l = 1$, the necessary condition for the absence of violations in the DS for each of the observations will have the form:

$$\beta_j^2 \in \chi^2(1, \ 2) \tag{27}$$

or by the rule of 3σ, for the quantile $a(1) = 0.02$,

$$\beta_j^2 \leq \gamma_1^2 = E\left[\beta_j^2\right] + 3\sqrt{V\left[\beta_j^2\right]} = 1 + 3\sqrt{2} \approx 5.2. \tag{28}$$

Based on reference data [4], quantiles $\tilde{a}(1) = 0.01$ and $\overline{a}(1) = 0.001$ can be matched to the extended $\tilde{\gamma}_1^2 \approx 7.6$ and $\overline{\gamma}_1^2 \approx 11.8$ tolerance values. However, when the tolerances are expanded, the probability of fault detection errors associated with a "false alarm" increases.

Taking into account the γ_1^2 tolerance, the monitoring of DS using the diagnostic parameter β_j^2 for the χ^2 criterion is reduced to checking the following conditions:

$$\left.\begin{array}{ll} \text{if } \beta_j^2 \leq \gamma_1^2, & \text{there are no faults in the } j\text{th channel of the DS;} \\ \text{if } \beta_j^2 > \gamma_1^2, & \text{there are faults in the } j\text{th channel of the DS.} \end{array}\right\} \tag{29}$$

Thus, the decomposition of the residual vector allows the formation of scalar goodness-of-fit criteria. The use of such criteria makes it possible to increase the depth of diagnosis.

4. Fault diagnosis by the ϑ^2 fisher criterion

The use of the χ^2 criterion makes it possible to detect current DS faults in real time. In practice, there is also a need to analyze the DS operation over a certain period of time.

Appropriate diagnostic parameters can be formed based on retrospective data. Technology (22)–(26) for sequential processing of observations allows the formation of scalar diagnostic parameters based on a sample of residuals on a sliding time interval. To do this, you can use the ergodic properties of the estimates obtained by the Kalman filter, the parameters of which are a priori adjusted for the proper operation of the DS. These parameters are the variances of the innovation sequences (22) in each of the l observation channels. The predicted value of the variance of the residual (22) at the ith moment of time is determined by the ratio (23), and its estimate $\hat{\alpha}_{j(i)}$ is based on a sample of N values, that is,

$$\hat{\alpha}^2_{j(i)} = \frac{1}{N-1} \sum_{k=i-N+1}^{i} \left[\nu_{j(k)} - \bar{\nu}_{j(i)}\right]^2; \qquad j = \overline{1, l}, \tag{30}$$

where $\bar{\nu}_{j(i)} = \frac{1}{N} \sum_{k=i-N+1}^{i} \nu_{j(k)}$ is an estimate of the mathematical expectation of the jth residual at the ith instant of time. Taking into account condition (8), the absence of faults can be associated with the desired value of the mathematical expectation of the residuals, namely, $\bar{\nu}_{j(i)} = 0$. As a parameter that characterizes the DS state on the time interval $T = [t_{i\text{-}N+1}, t_i]$, the ratio of the actual variance $\hat{\alpha}^2_j$ to its predicted value α^2_j can be accepted, that is,

$$F_j = \hat{\alpha}^2_j / \alpha^2_j. \tag{31}$$

It is known [4] that when the condition (32) presented below is satisfied, parameter (31) has ϑ^2 or Fisher distribution, namely,

$$F_j \in \vartheta^2(b, c), \tag{32}$$

$$\text{where } b = N/(N\text{–}2) \text{ and } c = 4N(N-1)/\left[(N-2)^2(N-4)\right] \tag{33}$$

are tabular values, respectively, of the mathematical expectation and variance for the F_j parameter.

Taking into account the 3σ rule and a = 0.02 quantile, condition (32) can be represented as

$$F_j \le \eta_1^2 = E[F_j] + 3\sqrt{V[F_j]} = b + 3\sqrt{c}. \tag{34}$$

When the tolerance η_1^2 is taken into account, the DS diagnosis according to the ϑ^2 criterion using the F_j parameter is reduced to checking the following conditions:

$$\text{if } F_j \le \eta_1^2, \tag{35}$$

there were no faults in the DS for the *j*th channel of observations on the time interval $T = [t_{i-N+1}, t_i]$;

$$\text{if } F_j > \eta_1^2, \tag{36}$$

there were faults in the DS for the *j*th channel of observations on the time interval *T*.

The procedures (35) and (36) complements the test (29) in order to increase the reliability of the diagnosis.

5. Fault diagnosis by the χ^2 and the ϑ^2 combined criterion

When diagnosing DS, the problem of detecting short-term outliers against a background of failures arises. If short-term outliers are recognized, then they can be parried without affecting the operation of the DS. The problem of recognizing the type of violations can be solved by combining the χ^2 and the ϑ^2 goodness-of-fit tests. Indeed, DS diagnosis based on the AA criterion makes it possible to immediately detect all violations, both random short-term outliers and failures. Indeed, DS diagnosis based on the χ^2 criterion makes it possible to immediately detect all violations, both random short-term outliers and failures. At the same time, DS diagnosis based on the ϑ^2 criterion is implemented using a set of residuals on a sliding time interval. Random short-term outliers that may appear on a sliding time interval when calculating the F_j parameter are averaged and, as a rule, do not affect to the diagnosis results of the ϑ^2 test. On the other hand, the appearance of gradual and sudden failures, which are characterized by constant deviations of residuals from their nominal values, lead to the F_j parameter going beyond the tolerance limits. Thus, if a violation in the *j*th observation was detected according to both criteria, then most likely a permanent failure occurred in the DS, and if only according to the χ^2 criterion, then it was a random short-term outlier.

From the considered properties of the combined criterion of consent, the following algorithm for detecting and parrying violations in the observed DS is as follows:

- If there is no discrepancy between the diagnostic parameters relative to the tolerances, the residuals are processed using EKF.

- Anomalous observation signals detected by the χ^2 criterion are excluded from processing or processed with certain robust confidence coefficients.

- Violations detected by both criteria χ^2 and ϑ^2 are countered by connecting a redundant channel.

The procedures presented above allow for the diagnosis of DS with a depth of up to the element of the observations vector. At the same time, in practice, it becomes necessary to perform diagnosis with a depth up to the element of the DS errors vector.

6. Fault diagnosis based on joint filtering and smoothing procedures

The diagnosis of DS with a depth up to the elements of the errors vector can be implemented on the basis of joint procedures for filtering observations in "forward" time and smoothing estimates in "backward" time. By analogy with expression (10), the diagnostic parameter, which reflects the deviations of error vector estimates from their predicted root-mean-square values, can be represented by a quadratic form:

$$J_i = \nu_{i/N}^T \Delta P_i^{-1} \nu_{i/N}, \tag{37}$$

where $\nu_{i/N} = \delta_{f(i)} - \delta_{s(i)} = \Phi_{i+1}^{-1}\hat{x}_{i+1/N} - \hat{x}_{i/i}$;
$\delta_{f(i)} = x_i - \hat{x}_{i/i}$; $\delta_{s(i)} = \Phi_{i+1}^{-1}(x_{i+1} - \hat{x}_{i+1/N})$;
$\Delta P_i = P_{i/i} + \Phi_i^{-1} P_{i+1/N} \Phi_i^{-T}$;
$\hat{x}_{i/i}$ and $\hat{x}_{i/N}$ are estimates of the DS error vector at the ith moment of time, obtained, respectively, at the filtering and smoothing stage; $P_{i/i}, P_{i/N}$ are the covariance matrices of the above estimates; $\Phi^{-T} = (\Phi^{-1})^T$. Φ_i^{-1} is the inverse transition matrix, which is determined from the solution of the differential equation:

$$d\Phi^{-1}(t, t_{i-1})/dt = -\Phi^{-1}(t, t_{i-1})A(t);\ \Phi^{-1}(t_{i-1}, t_{i-1}) = I. \tag{38}$$

Stable smoothing ($|\delta| < 3\sigma$), which reflects the absence of anomalous observations signals, is characterized by the following distribution of the residual $\nu_{i/N}$ and the quadratic form $J_{s(i)}$:

$$\nu_{i/N} \in N(0; \Delta P_i);\ J_{s(i)} \in \chi^2(n; 2n), \tag{39}$$

where n is the dimension of the DS error vector.

Taking into account the properties of the χ^2 distribution and the 3σ rule, it is possible to formulate the necessary conditions for the absence of violations for the DS in general and for the jth element of the error vector in particular

$$J_{s(i)} \leq n + 3\sqrt{2n}; \tag{40}$$

$$J_{s(i/j)} = J_{s(i/j-1)} + \tilde{\nu}_{i/N(j)}^2 / \Delta D_{i(j)} \leq \gamma_j^2 = j + 3\sqrt{2j}, \tag{41}$$

where $\tilde{\nu}_{i/N} = \Delta U_i^{-1} \nu_{i/N}$; $j = \overline{1, n}$; ΔU_i^{-1}; ΔD_i^{-1} are an upper triangular matrix with identity diagonal and a diagonal matrix, respectively. These matrices are obtained by the following orthogonal transformation:

$$\Delta P_i^{-1} = \Delta U_i^{-T} \Delta D_i^{-1} \Delta U_i^{-1}; \tag{42}$$

$\Delta D_{i(j)}^{-1}$ is the jth element of the diagonal matrix ΔD_i^{-1}.

Taking into account decomposition (42) and the statistical properties of the Fisher distribution [4]:

$$F_{s(j)} = \hat{\alpha}_{(i/j)} / \Delta D_{(i/j)} \in \vartheta^2(b, c), \tag{43}$$

it is possible to formulate the necessary condition for the operable DS state ("absence" of failures) for the jth element of the error vector, that is,

$$F_{s(j)} \leq \eta_j^2 = b + 3\sqrt{c}, \tag{44}$$

where $\hat{\alpha}_{(i/j)}$ is an estimate of the variance of the residual $\tilde{\nu}_{i/N(j)}$ on a sliding time interval; b and c are tabulated values (33) of the mathematical expectation and variance for the parameter F_j.

Diagnosis based on joint procedures for optimal filtering and smoothing assumes the relationship between prediction and interpolation models of the DS error vector:

$$x_{i+1} = \Phi_{i+1}x_i + \Gamma_{i+1}\xi_i \text{ prediction one step ahead at } i = \overline{i_0, i_f}; \tag{45}$$

$$x_i = \Phi^{-1}{}_{i+1}(x_{i+1} - \Gamma_{i+1}\xi_i) \text{ interpolation one step back at } i = \overline{i_f, i_0}. \tag{46}$$

It should be noted, however, that the traditional RTS (Rauch-Tung-Striebel) algorithm [9] smooths out the predicted estimates $\hat{x}_{i/i-1}$, the reliability of which is significantly lower than the updated estimates $\hat{x}_{i/i}$.

The value $\hat{x}_{(i/i)N}$ for the smoothed estimate of the x_i error vector can be found by solving the following optimization problem using the least squares method

$$\hat{x}_{(i/i)N} = \arg\min_{x_i} \; J_{i/s}, \tag{47}$$

$$\text{where} \quad J_{i/s} = 0.5\Big\{\underbrace{\left[x_i - \hat{x}_{(i/i+1)N}\right]^T P^{-1}_{(i/i+1)N}\left[x_i - \hat{x}_{(i/i+1)N}\right]}_{a} + \underbrace{\left[x_i - \hat{x}_{i/i}\right]^T P^{-1}_{i/i}\left[x_i - \hat{x}_{i/i}\right]}_{b}\Big\}; \tag{48}$$

$\hat{x}_{(i/i+1)N}$; $P_{(i/i+1)N}$ is an estimate of the error vector interpolated at the ith moment of time and its covariance matrix.

The quadratic form (48) combines components "a" and "b", which characterize interpolation and filtering errors at the ith moment of time. It should be noted that component "b" can also be formed for predicted estimates. However, in this case, smoothing errors will significantly depend on the frequency of registration these estimates.

Applying the rule of differentiation with respect to the parameter for the quadratic form (48), we can obtain the following solution to problem (47)

$$\left.\frac{\partial J_i}{\partial x_i^T}\right|_{x_i = \hat{x}_{(i/i)N}} = P^{-1}_{(i/i+1)N}\left[\hat{x}_{(i/i)N} - \hat{x}_{(i/i+1)N}\right] + P^{-1}_{i/i}\left[\hat{x}_{(i/i)N} - \hat{x}_{i/i}\right] = 0. \tag{49}$$

From (Eq. (49)), an expression can be obtained for a smoothed estimate of the error vector

$$\hat{x}_{(i/i)N} = \hat{x}_{(i/i+1)N} - P_{(i/i)N}P^{-1}_{i/i}\left[\hat{x}_{(i/i+1)N} - \hat{x}_{i/i}\right], \tag{50}$$

$$\text{where } P_{(i/i)N} = \left[P^{-1}_{(i/i+1)N} + P^{-1}_{i/i}\right]^{-1}; \tag{51}$$

$$K_{i/N} = P_{(i/i)N}P^{-1}_{i/i} \text{ is the optimal amplification coefficient.} \tag{52}$$

Applying the matrix inversion lemma [10] to expression (51), it can be represented in expanded form:

$$P_{(i/i)N} = P_{(i/i+1)N} - P_{(i/i+1)N}\left[P_{(i/i+1)N} + P_{i/i}\right]^{-1}P_{(i/i+1)N}. \tag{53}$$

Substituting expression (53) into relation (52), we obtain the following procedure for calculating the optimal amplification coefficient

$$K_{i/N} = P_{(i/i+1)N}\left[P_{(i/i+1)N} + P_{i/i}\right]^{-1}. \tag{54}$$

The expression for the interpolated estimate follows from the ratio (46)

$$\hat{x}_{(i/i+1)N} = \Phi_i^{-1}\hat{x}_{(i+1/i+1)N}. \tag{55}$$

The difference between relations (46) and (55) determines the equation for interpolation errors

$$e_{(i/i+1)N} = \Phi_i^{-1}\left[e_{(i+1/i+1)N} - \Gamma_i\xi_i\right], \tag{56}$$

where $e_{(i/i+1)N} = x_i - \hat{x}_{(i/i+1)N}$; $e_{(i+1/i+1)N} = x_{i+1} - \hat{x}_{(i+1/i+1)N}$.
The following covariance matrix corresponds to the interpolation error vector

$$P_{(i/i+1)N} = E\left[e_{(i/i+1)N}e^T_{(i/i+1)N}\right] = \Phi_i^{-1}\left[P_{(i+1/i+1)N} + \Gamma_i Q_i \Gamma_i^T\right]\Phi_i^{-T}. \tag{57}$$

Combining (Eqs. (50), (53)–(55), (57)) into a single structure, we obtain an algorithm for optimal smoothing of error vector estimates
Interpolation:

$$\hat{x}_{(i/i+1)N} = \Phi_{i+1}^{-1}\hat{x}_{(i+1/i+1)N}; \tag{58}$$

$$\tilde{P}_{(i/i+1)N} = P_{(i+1/i+1)N} + \Gamma_{i+1}Q_i\Gamma_{i+1}^T. \tag{59}$$

Updating:

$$\nu_{i/N} = \hat{x}_{(i/i+1)N} - \hat{x}_{i/i}; \tag{60}$$

$$P^{-1}_{(i/i)N} = \Phi^T_{i+1}\tilde{P}^{-1}_{(i/i+1)N}\Phi_{i+1} + P^{-1}_{i/i}; \tag{61}$$

$$K_{i/N} = P_{(i/i)N}P^{-1}_{i/i}; \tag{62}$$

$$\hat{x}_{(i/i)N} = \hat{x}_{(i/i+1)N} - K_{i/N}\nu_{i/N}, \tag{63}$$

To implement the algorithm (58)–(63), it is necessary to register estimates $\hat{x}_{i/i}$of the DS error vector and their covariance matrices $P^{-1}_{i/i}$ obtained by the EKF in the process of functional monitoring of the DS. The matrix $P^{-1}_{i/i}$ is determined from the solution in “forward” time according to the following equations:

$$P^{-1}_{i/i} = P^{-1}_{i/i-1} + H_i^T R_i^{-1} H_i; \tag{64}$$

$$P_{i/i-1}^{-1} = \left(\Phi_i P_{i-1/i-1} \Phi_i^T + \Gamma_i Q_{i-1} \Gamma_i^T\right)^{-1}. \tag{65}$$

The implementation of (Eqs. (58)–(65)) is associated with computational difficulties due to the need to determine the inverse covariance matrices $P_{i/i}^{-1}$ and $P_{(i/i)N}^{-1}$. These difficulties can be significantly reduced with the sequential implementation of smoothing algorithms. With this implementation, the inversion of covariance matrices of $n \times n$ dimension is replaced by a sequence of division operations. Such a replacement is based on the transformation of (Eq. (61), (64), (65)) according to the matrix inversion lemma [10]. For (Eq. (65)) it will have the form:

$$P_{i/i-1}^{-1} = \tilde{P}_{i/i-1}^{-1} - \tilde{P}_{i/i-1}^{-1} \Gamma_i \left(\Gamma_i^T \tilde{P}_{i/i-1}^{-1} \Gamma_i + Q_{i-1/i-1}^{-1}\right)^{-1} \Gamma_i^T \tilde{P}_{i/i-1}^{-1} = \left(\tilde{P}_{i/i-1} + \sum_{j=i}^{r} \Gamma_j \Gamma_j^T Q_{jj}\right)^{-1}, \tag{66}$$

where Γ_j is the jth column of matrix Γ_i, which has $n \times r$ dimension; Q_{jj} is the jth element of the diagonal matrix Q_i, which has $r \times r$ dimension.

Hence, the following procedure for the sequential implementation of Eq. (66) is valid

$$M_0^{-1} = \Phi_i^{-T} P_{i-1/i-1}^{-1} \Phi_i^{-1}; \tag{67}$$

$$K_j = M_{j-1}^{-1} \Gamma_j / \left(\Gamma_j^T M_{j-1}^{-1} \Gamma_j + Q_{jj}^{-1}\right); \tag{68}$$

$$M_j^{-1} = \left(K_j \Gamma_j^T - I\right) M_{j-1}^{-1} \left(K_j \Gamma_j^T - I\right)^T + K_j Q_{jj}^{-1} K_j^T; \tag{69}$$

$$P_{i/i-1}^{-1} = M_r^{-1}; \; j = \overline{1, r}. \tag{70}$$

Transformations similar to (66)–(70) will also be valid for Eq. (61), if the covariance matrix of filtering errors is presented in the following form

$$P_{i/i}^{-1} = U_{i/i}^{-T} D_{i/i}^{-1} U_{i/i}^{-1}, \tag{71}$$

where $U_{i/i}$ is an upper triangular matrix with unit diagonal elements; $D_{i/i}$ is a diagonal matrix. These matrices form the basis of the *U-D* filter [11].

Taking into account decomposition (71) and preliminary decomposition of residuals $\tilde{\nu}_{i/N} = U_i^{-1} \nu_{i/N}$, a sequential modification of the smoothing and diagnostic algorithm can be formed.

Interpolation:

$$m_0 = \tilde{x}_{(i/i+1)N} = \Phi_{i+1}^{-1} \hat{x}_{(i+1/i+1)N}; \tag{72}$$

$$\nu_{i/N} = \tilde{x}_{(i/i+1)N} - \hat{x}_{i/i}; \tag{73}$$

$$MWGS \left\{ \begin{array}{c} \overline{W}_0 = \left[\Phi_{i+1}^{-1} U_{i+1/N} \vdots \Phi_{i+1}^{-1} \Gamma_{i+1}\right] \\ \overline{D}_0 = diag\left(D_{i+1/N}, Q_i\right) \end{array} \right\} \rightarrow \begin{array}{c} \tilde{U}_0 \\ \tilde{D}_0 \end{array}; \tag{74}$$

Smoothing:

$$f_j = \overline{U}_j^{-1}\,\tilde{U}_{j-1};\; V_j = \tilde{D}_j f_j^T; \tag{75}$$

$$\alpha_j^2 = f_j V_j + D_{jj}; \tag{76}$$

$$\tilde{K}_j = \tilde{U}_{j-1}\, V_j/\alpha_j^2; \tag{77}$$

$$MWGS\left\{\begin{array}{c} \overline{W}_j = \left[\left(\tilde{K}_j f_j - \tilde{U}_{j-1}\right) \vdots \tilde{K}_j\right] \\ \overline{D}_j = diag\left(\tilde{D}_{j-1}; D_{jj}\right) \end{array}\right\} \rightarrow \begin{array}{c}\tilde{U}_j \\ \tilde{D}_j\end{array}; \tag{78}$$

$$K_j = \tilde{U}_j\, \tilde{D}_j\, \tilde{U}_j^T\, \overline{U}_j^{-T} D_j^{-1}; \tag{79}$$

$$\tilde{\nu}_{i/N(j)} = \overline{U}_j^{-1}\, \nu_{i/N}; \tag{80}$$

$$m_j = m_{j-1} - K_j\, \tilde{\nu}_j;\; j = \overline{1, n}; \tag{81}$$

$$\hat{x}_{i/N} = m_n; U_{(i/i)N} = \tilde{U}_n; D_{(i/i)N} = \tilde{D}_n, \tag{82}$$

where $\overline{U}_j^{-1}$ is the jth row of the matrix $U_{i/i}^{-1}$; D_j^{-1} is the jth element of the diagonal matrix $D_{i/i}^{-1}$; MWGS is the modified weighted Gram-Schmidt procedure [11], intended to transform the aggregate of matrices $\overline{W}_j$ and $\overline{D}_j$, which are $n \times (n + r)$ matrix and $(n + r) \times (n + r)$ matrix, respectively, into the aggregate of the $n \times n$ an upper triangular matrix $\tilde{U}_j$ with identity diagonal and a diagonal matrix $\tilde{D}_j$. The algorithm for implementing the MWGS procedure has the form

$$d_m = \sum_{j=1}^{n+r} \overline{W}_{mj} \overline{D}_j \overline{W}_{mj}, \quad m = \overline{n, 2}; \tag{83}$$

$$\left.\begin{array}{l} u_{km} = \sum\limits_{j=1}^{n+r} \overline{W}_{mj} \overline{D}_j \overline{W}_{mj}/d_m, \\ \overline{W}_{kj} := \overline{W}_{kj} - u_{km} \overline{W}_{mj},\;\; j = \overline{1, n + r} \end{array}\right\} \quad k = \overline{1, m - 1}; \tag{84}$$

$$d_1 = \sum_{j=1}^{n+r} \overline{W}_{1j} \overline{D}_j \overline{W}_{1j}; \tag{85}$$

u_{km} and d_j are the elements of the required matrix U and D.

To implement the *U-D* smoothing algorithm, it is necessary to determine the $U_{i/i}^{-1}$, $D_{i/i}^{-1}$ components of the a posteriori covariance matrix of estimation errors $P_{i/i}^{-1}$. The traditional *U-D* algorithm [11] for observation filtering, which includes additional procedures necessary to smoothing estimates of the DS error vector, is presented below.

Prediction for filtering:

$$m_0 = x_{i/i-1} = \Phi_i \hat{x}_{i-1/i-1}; \tag{86}$$

$$MWGS \left\{\begin{array}{c} \overline{W}_i = \left[\Phi_i\, U_{i-1/i-1} \vdots \Gamma_i\right] \\ \overline{D}_i = diag\;\left(D_{i-1/i-1}, Q_{i-1}\right) \end{array}\right\} \rightarrow U_0; D_0. \tag{87}$$

Prediction for smoothing:

$$U_0^{-T} = \Phi_i^{-T} U_{i-1/i-1}^{-T}; D_0^{-1} := D_{i-1/i-1}^{-1}; \tag{88}$$

$$f_j = \Gamma_j \, U_{j-1}^{-T}; \quad V_j = D_{j-1}^{-1} f_j^T; \tag{89}$$

$$K_j = U_{j-1}^{-T} \, V_j / \left(f_j V_j + Q_{jj}^{-1} \right); \tag{90}$$

$$MWGSL \left\{ \begin{array}{l} \tilde{W}_j = \left[K_j f_j - U_j^{-T} \vdots K_j \right] \\ \tilde{D}_j = \; diag\left(D_j^{-1}, Q_{jj}^{-1} \right) \end{array} \right\} \to U_j^{-T}; D_j^{-1}; \; j = \overline{1, r}; \tag{91}$$

$$U_0^{-T} := U_{i/i-1}^{-T} = U_r^{-T}; D_0^{-1} := D_{i/i-1}^{-1} = D_r^{-1}. \tag{92}$$

Filtering:

$$f_j = H_j U_{j-1}; \quad V_j = D_{j-1} f_j^T; \tag{93}$$

$$\alpha_j^2 = f_j V_j + R_j; \tag{94}$$

$$K_j = U_{j-1} V_j / \alpha_j^2; \tag{95}$$

$$m_j = m_{j-1} + K_j \nu_j; \tag{96}$$

$$MWGS \left\{ \begin{array}{l} \overline{W}_j = \left[K_j f_j - U_{j-1} \vdots K_j \right] \\ \overline{D}_j = diag(D_{j-1}, R_j) \end{array} \right\} \to U_j; D_j. \; j = \overline{1, l}. \tag{97}$$

Updating for smoothing:

$$MWGSL \left\{ \begin{array}{l} \tilde{W}_j = \left[U_{j-1}^{-T} \vdots H_j^T \right] \\ \tilde{D}_j = diag\left(D_{j-1}^{-1}, R_j^{-1} \right) \end{array} \right\} \to U_j^{-T}; D_j^{-1}; \; j = \overline{1, l}; \tag{98}$$

$$U_{i/i}^{-T} = U_l^{-T}; D_{i/i}^{-1} = D_l^{-1}, \tag{99}$$

where MWGSL is a similar procedure to *MWGS* for forming matrices: the lower triangular with unit diagonal elements U_j^{-T} and diagonal D_j^{-1}. Eq. (88)–(92), (98), (99) are redundant with respect to the basic *U-D* filtering algorithm [11].

Diagnosis in "backward time" is performed when violations detected in "forward time".

7. Analysis of the results of mathematical modeling of diagnosis procedures based on filtering and smoothing of observations

The Schuler pendulum [12], which includes a gyro-accelerometer (G-A) system moving in a gravitational field along a sphere of R radius, is considered as an object of mathematical modeling. The invariance of the modeled vertical (radius vector) to the movement of the base of the G-A system along the sphere when changing the velocity V and angular position φ is ensured by precession of the gyros. For this purpose, a $\dot{\varphi} = V/R$ signal that is proportional to angular velocity of G-A system motion relative to the sphere is fed to the gyros moment sensor. When the vector $x(t) = [\Delta V \; \delta \; \Delta a \; \Delta\omega]^{\mathrm{T}}$

of errors of such a system is observed from the velocity signals $Z_V(t) = V_{\text{G-A}}(t) - V_{\text{SEI}}(t)$ the parameters of Eqs. (3) and (5) will have the following form

$$A(t) = \begin{bmatrix} 0 & -g & 1 & 0 \\ 1/R & 0 & 0 & 1 \\ 0 & 0 & -1/\tau_a & 0 \\ 0 & 0 & 0 & -1/\tau_\omega \end{bmatrix}; \qquad (100)$$

$$G(t) = \begin{bmatrix} 0 & 0 & 0 & 0 \\ 0 & 0 & 0 & 0 \\ 0 & 0 & \sigma_a\sqrt{2/\tau_a} & 0 \\ 0 & 0 & 0 & \sigma_\omega\sqrt{2/\tau_\omega} \end{bmatrix}; \qquad (101)$$

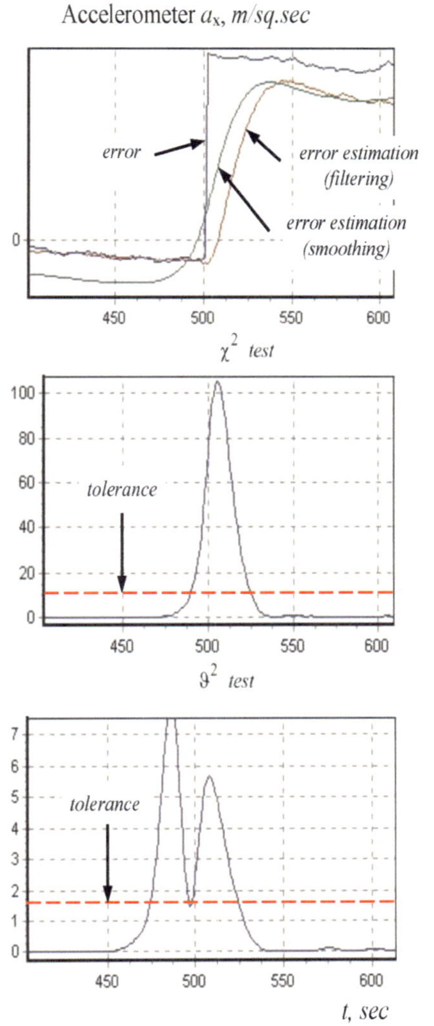

Figure 2.
Estimates of the accelerometer error.

$$H(t) = [1\ 0\ 0\ 0]^{\mathrm{T}}, \tag{102}$$

where g is the acceleration of gravity; δ is an angular error of determining the vertical; ΔV is an error of velocity reckoning; Δa is an accelerometer error; $\Delta\omega$ is the gyro drift; τ_a, τ_ω are the correlation time for the accelerometer error and gyro drift, respectively; σ_a, σ_ω are mean-square values of errors of the accelerometer and gyro, respectively; $\Delta(\ldots)$ stands for an error.

In **Figures 2** and **3**, characteristic results of the studies of an algorithm for G-A system diagnosis from the recorded data are presented. At the 500th second, the accelerometer failure corresponding to the imitation bias was simulated. Such a failure indirectly makes itself evident during filtering for the velocity observation channel when the diagnostic parameter β_V^2 is above upper tolerance.

In post processing the recorded estimates and during the diagnosis by rules (41) and (44), it is possible to determine which of the sensors – an accelerometer or a gyro – has most probably caused a fault.

Figures 2 and **3** show the dynamics of variation of the estimates of both the bias of the a_x accelerometer output signal and the ω_x gyro drift when velocity observations are processed in "forward time" and when the estimates mentioned are refined in "backward time". During the diagnosis from the recorded data, a failed accelerometer is isolated if the generalized parameters J_{Sa_j} (the χ^2 test) and F_{Sa_j} (the ϑ^2 test) are above upper tolerances (see **Figure 2**). It can also be seen (see **Figure 3**) that the accelerometer failure has had an insignificant influence on the

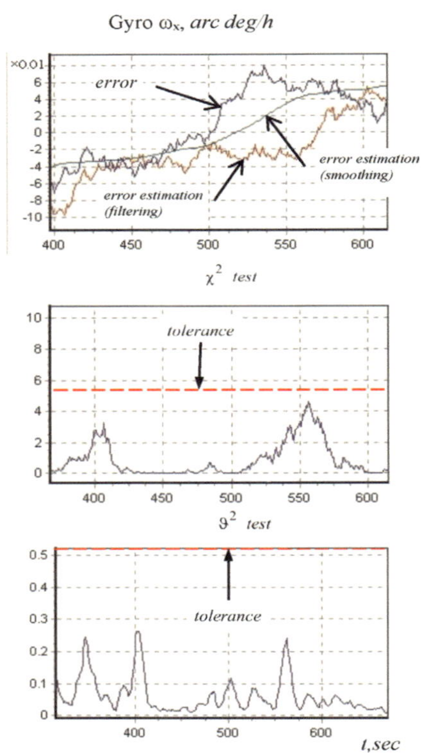

Figure 3.
Estimates of the gyro drift.

variation of the generalized parameters $J_{S\omega_j}$ and $F_{S\omega_j}$, which characterize the status of the ω_x gyro.

Thus, combined processing of observations in "forward" and "backward" time enables us to solve diagnosis problems with a depth of the component of the state vector of a dynamical system.

8. Analysis and parrying of violations by robust filtering

The analysis and parrying of anomalous observations can be performed based on a robust modification of the *U-D* filter. This modification relies on the application of the influence function ψ, which defines the level of confidence in incoming observations. Such a function can be formed for the normalized residual $\beta_j = \nu_j/\alpha_j$, where α_j is a scaling parameter which may be the mean square deviation $\sqrt{R_j}$. For the residual β_j, one can not only form the robust-likelihood function $\rho(\beta)$ but also perform the optimization of the estimates

$$\hat{x}_i = \arg\min_{x_i} \sum_{i=i_0+1}^{i_f} \rho(\beta_i), \tag{103}$$

where $\rho(\beta) = -\ln f(\beta); f(\beta)$ is a probability density function (PDF) [3].

The solution of the problem (116), in view of the constraint $x_i - \Phi_i x_{i-1} - \Gamma_i \xi_{i-1} = 0$, is an algorithm for robust estimation.

Prediction: Eq. (97), (98)

Tuning:

$$\beta_j = \nu_j/\alpha_j; \quad \psi_j = \psi\left(\beta_j\right); \quad \psi'_j = \psi'\left(\beta_j\right); \tag{104}$$

Updating:

$$f_j = H_j U_{j-1}; \quad V_j = D_{j-1} f_j^T; \tag{105}$$

$$\tilde{\alpha}_j^2 = f_j V_j \, \psi' + \alpha_j^2; \tag{106}$$

$$K_j = U_{j-1} V_j / \tilde{\alpha}_j^2; \tag{107}$$

$$m_j = m_{j-1} + K_j \alpha_j \psi_j; \tag{108}$$

$$MWGS \left\{ \begin{array}{l} \overline{W}_j = \left[K_j f_j \, \psi'_j - U_{j-1} \vdots K_j\right] \\ \overline{D}_j = diag\left(D_{j-1}, \alpha_j^2 \psi'_j\right) \end{array} \right\} \to U_j; D_j; \tag{109}$$

$$U_{i/i} = U_l; D_{i/i} = D_l; \hat{x}_{i/i} = m_l; j = \overline{1, l}, \tag{110}$$

where $\psi_j = \psi\left(\beta_j\right) = \frac{\partial \rho\ (\beta)}{\partial \beta} \Big|_{\beta = \beta_j}$ and $\psi'_j = \psi'\left(\beta_j\right) = \frac{\partial^2 \rho\ (\beta)}{\partial \beta^2} \Big|_{\beta = \beta_j}$.

The influence function [13, 14] ψ_j and its derivative ψ'_j can be formed taking into account "a priori" assumptions about the laws of distribution of the useful signal and noise. The choice of the values of above functions is based on the necessary conditions

[15] for the filter to be divergence-free, namely, the diagnostic parameter must have a Gaussian distribution, that is, $\beta \in N(0,1)$; in addition, the rule of 3σ is fulfilled [3] for the probability P that a random variable having the Gaussian distribution will be on the interval $[-3\sigma, 3\sigma]$, that is,

$$P\{|\beta - E[\beta]| \geq 3\sigma\} = 0.0027. \tag{111}$$

For the random variable β, the rule (111) can be written as $P\{|\beta| < 3\} = 0.9973$.

Thus, the correct functioning of the DS can be matched by the condition $|\beta| < 3$ and also the following values of the functions:

$$f_g(\beta) = (2\pi)^{-\frac{1}{2}} \exp(-0.5\beta^2); \ \ \rho_g(\beta) = 0.5\ln(2\pi) + 0.5\beta^2; \ \ \psi_g(\beta_j) = \beta_j; \ \ \psi_g'(\beta_j) = 1. \tag{112}$$

The following functions can be made to correspond to such a distribution and to off-design conditions of the filter operation.

$$f_l(\beta) = 0.5\exp(-|\beta|); \ \ \rho_l(\beta) = \ln 2 + |\beta|; \ \ \psi_l(\beta_j) = |1|; \ \ \psi_l'(\beta_j) = 0. \tag{113}$$

The vagueness of boundaries between anomalous and conditioned signals can be taken into account based on the convolution of the Gauss and Laplace PDF. Such a convolution can be performed using the following moment generating functions (MGFs) [16]:

$$M_{lg}(T) = M_l(T)M_g(T) = \left(1 - T^2\right)^{-1} \exp\left(T^2/2\right), \tag{114}$$

where $M(T)$ is an MGF; T is generally a complex number; and $K(T) = \ln M(T)$ is a cumulant generating function.

Relying on the results of [14], the following relations can be shown to hold for the normalized residual β_j:

$$\psi(\beta_j) = T_0 + K^{(3)}(T)/2|_{T=T_0}; \tag{115}$$

$$\psi'(\beta_j) = \left[1 + \partial K^{(3)}(T)/\partial T\right] \left.\frac{\partial T}{\partial \beta}\right|_{T=T_0}, \tag{116}$$

where T_0 is the value of the argument at a saddle point for which the following equality is valid:

$$K_T'(T) - \beta_j = 0. \tag{117}$$

In view of the approximation $\ln(1 - T^2) \approx -T^2$ and relations (112)–(117), the parameters $\psi_j$and ψ_j' take the form $\psi_{lg}(\beta_j) = \beta/3$; $\psi_{lg}'(\beta_j) = 1/3$.

The influence function which reflect the considered assumptions is shown in **Figure 4**.

In this figure, a typical influence function is denoted by a dotted line, too. However, the interrelation between such a function and the parameter ψ_j' is not obvious.

The need to parry anomalous observations can be illustrated by the results of mathematical modeling of a G-A system, which is described by Eq. (100)–(102). Such

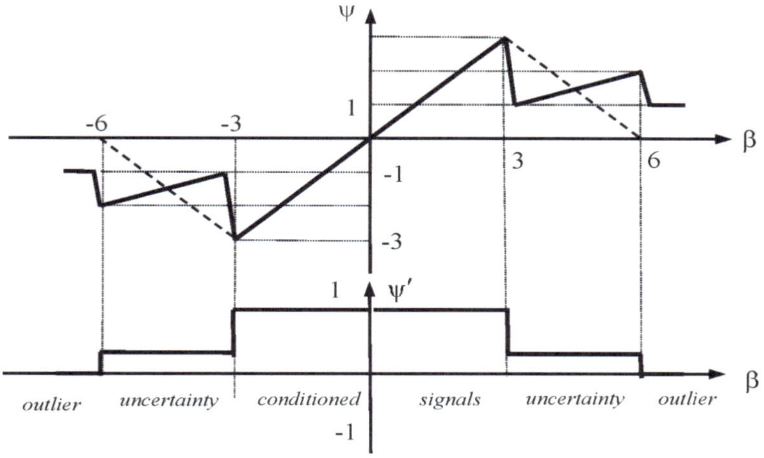

Figure 4.
Diagram for the control of an estimating filter with an influence function.

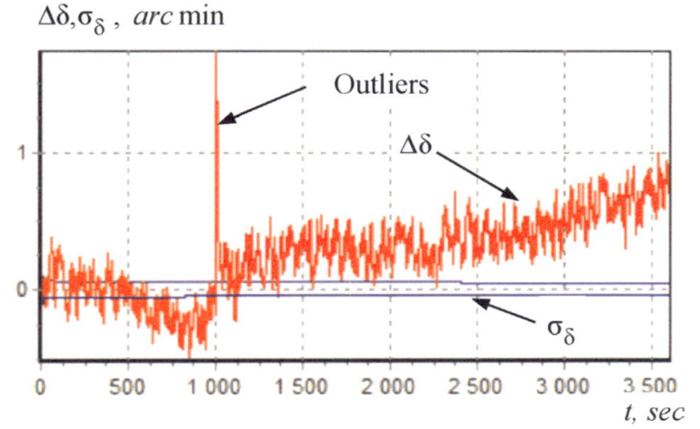

Figure 5.
Estimates obtained by the EKF.

results are presented in **Figures 5** and **6**, where the true $\Delta\delta = x_\delta - \hat{x}_\delta$ and root mean square (RMS) error $\sigma_\delta = \sqrt{P_\delta}$ of vertical determination are shown. An anomalous observation signal $Z_V(t)$ was simulated. In the case of a sequential EKF (20)–(26), estimation errors are shown in **Figure 5**, and for a robust filter (97)–(98) and (100)–(117), such errors are presented in **Figure 6**.

It can be seen that in the absence of protection from anomalous observations, the divergence of the EKF is possible when the actual estimation errors $\Delta\delta$ considerably differ from their predicted mean square values σ_δ.

9. Conclusions

In this chapter, some approaches to monitoring and diagnosis of dynamic systems using the extended Kalman filter were presented. The considered approaches are based on the application of the DS error model, the formation and processing of

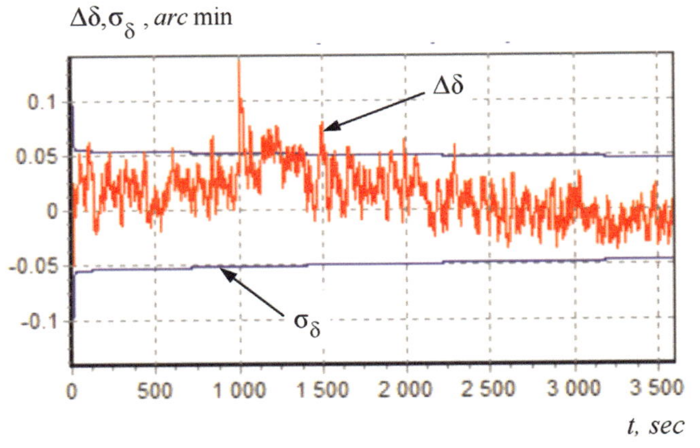

Figure 6.
Estimates obtained by the robust filter.

residuals between the observed parameters and their predicted values. The statistical properties of residuals make it possible to form diagnostic parameters and the corresponding criteria of agreement. For the considered criteria, formalized tolerances can be generated that make it possible to detect failures and failures in the DS.

The conducted studies have confirmed the following possibilities of the proposed approaches.

- Detection and exclusion or processing with certain confidence coefficients of anomalous observations.
- Selection of outliers against the background of gradual biases of observations.
- Improving the reliability of monitoring based on combined goodness-of-fit tests.
- Detection of faults in observable dynamical systems with a depth of the component of the state vector on a basis of the joint procedures of optimal filtering and smoothing of experimental data.

Author details

Alexander Chernodarov
MAI University, Moscow, Russia

*Address all correspondence to: chernod@mail.ru

References

[1] Maybeck P. Stochastic Models, Estimation and Control. Vol. 2. New York: Academic Press; 1982

[2] Gertler J. Fault Detection and Diagnosis in Engineering Systems. New York: Marcel Dekker; 1998

[3] Skorokhod A, Hoppensteadt F, Salehi H. Random Perturbation Methods with Applications in Science and Engineering. New York: Springer-Verlag; 2002

[4] Korn G, Korn T. Mathematical Handbook. New York: Mc Graw-Hill; 1968

[5] Tanil C, Khanafseh S, Joerger M, et al. An INS monitor to detect GNSS spoofers capable of tracking vehicle position. IEEE Transactions on Aerospace and Electronic Systems. 2018;**54**(1):131-143

[6] Yu Z, Zhang Q, Yu K, Zheng N. A state-domain robust Chi-Square test method for GNSS/INS integrated navigation. Journal of Sensors. 2021; **2021**:1745383, 8 pages. DOI: 10.1155/2021/1745383SINS

[7] Wilkinson J, Reinsch C. Handbook for Automatic Computation. Vol. II, Linear Algebra. Berlin: Springer-Verlag; 1971

[8] Bierman G. Factorization Methods for Discrete Sequential Estimation. New York: Academic Press; 1977

[9] Rauch H, Tung F, Striebel C. Maximum likelihood estimates of linear dynamic systems. AIAA Journal. 1965;**3**(8):1445-1450

[10] Sage A, White C. Optimum Systems Control. New Jersey: Prentice – Hall; 1977

[11] Thornton C, Bierman G. UDU^T covariance factorization for Kalman filtering. In: Control and Dynamic Systems. New York: Academic Press; 1980. pp. 117-247

[12] Rogers R. Applied Mathematics in Integrated Navigation Systems. 2nd ed. Reston: AIAA Education Series; 2003

[13] Abutaleb A, Papaiouannou M. New results in Sridhar filtering theory: The discrete case. Journal of Optimization Theory and Applications. 1990;**64**(1):5-14

[14] Hampel F, Ronchetti E, Rousseeuw P. Robust Statistics: The Approach Based on Influence Functions. New York: Wiley; 1986

[15] Fitzgerald R. Divergence of the Kalman filter. IEEE Transactions on Automatic Control. 1971;**16**(6):736-747

[16] Wu W. Target tracking with glint noise. IEEE Transactions on Aerospace and Electronic Systems. 1993;**29**(1):174-185